T0296951

LONDON MATHEMATICAL SOCIETY LECTURE NOTE SERIES

Managing Editor: Professor M. Reid, Mathematics Institute, University of Warwick, Coventry CV4 7AL, United Kingdom

The titles below are available from booksellers, or from Cambridge University Press at
www.cambridge.org/mathematics

London Mathematical Society Lecture Note Series: 386

Independence-Friendly Logic
A Game-Theoretic Approach

ALLEN L. MANN
University of Tampere, Finland

GABRIEL SANDU
University of Helsinki, Finland

MERLIJN SEVENSTER
Philips Research Laboratories, The Netherlands

CAMBRIDGE
UNIVERSITY PRESS

CAMBRIDGE
UNIVERSITY PRESS

University Printing House, Cambridge CB2 8BS, United Kingdom

One Liberty Plaza, 20th Floor, New York, NY 10006, USA

477 Williamstown Road, Port Melbourne, VIC 3207, Australia

314-321, 3rd Floor, Plot 3, Splendor Forum, Jasola District Centre, New Delhi - 110025, India

103 Penang Road, #05-06/07, Visioncrest Commercial, Singapore 238467

Cambridge University Press is part of the University of Cambridge.

It furthers the University's mission by disseminating knowledge in the pursuit of
education, learning and research at the highest international levels of excellence.

www.cambridge.org
Information on this title: www.cambridge.org/9780521149341

© A. L. Mann, G. Sandu and M. Sevenster 2011

First published 2011

A catalogue record for this publication is available from the British Library

ISBN 978-0-521-14934-1 Paperback

Contents

1

Introduction

First-order logic meets game theory as soon as one considers sentences with alternating quantifiers. Even the simplest alternating pattern illustrates this claim:

$$\forall x \exists y (x < y). \tag{1.1}$$

We can convince an imaginary opponent that this sentence is true on the natural numbers by pointing out that for every natural number m he chooses for x, we can find a natural number n for y that is greater than m. If, on the other hand, he were somehow able to produce a natural number for which we could not find a greater one, then the sentence would be false.

We can make a similar arrangement with our opponent if we play on any other structure. For example, if we only consider the Boolean values 0 and 1 ordered in their natural way, we would agree on a similar protocol for testing the sentence, except that each party would pick 0 or 1 instead of any natural number.

It is natural to think of these protocols as *games*. Given a first-order sentence such as (1.1), one player tries to verify the sentence by choosing a value of the existentially quantified variable y, while the other player attempts to falsify it by picking the value of the universally quantified variable x. Throughout this book we will invite Eloise to play the role of verifier and Abelard to play the role of falsifier.

We can formalize this game by drawing on the classical theory of extensive games. In this framework, the game between Abelard and Eloise that tests the truth of (1.1) is modeled as a two-stage game. First Abelard picks an object m. Then Eloise observes which object Abelard chose, and picks another object n. If $m < n$, we declare that Eloise has won the game; otherwise we declare Abelard the winner. We notice that

Eloise's ability to "see" the object m before she moves gives her an advantage. The reason we give Eloise this advantage is that the quantifier $\exists y$ lies within the scope of $\forall x$. In other words, the value of y depends on the value of x.

Hintikka used the game-theoretic interpretation of first-order logic to emphasize the distinction between constitutive rules and strategic principles [28, 29]. The former apply to individual moves, and determine whether a particular move is correct or incorrect. In other words, constitutive rules determine the set of all possible *plays*, i.e., the possible sequences of moves that might arise during the game. In contrast, strategic principles pertain to the observed behavior of the players over many plays of the game. Choosing blindly is one thing, following a strategy is another. A strategy is a rule that tells a particular player how to move in every position where it is that player's turn. A winning strategy is one that ensures a win for its owner, regardless of the behavior of the other player(s). Put another way, constitutive rules tell us how to play the game, while strategic principles tell us how to play the game well.

When working with extensive games, it is essential to distinguish between winning a single play, and having a winning strategy for the game. If we are trying to show that (1.1) holds, it is not enough to exhibit one single play in which $m = 4$ and $n = 7$. Rather, to show (1.1) is true, Eloise must have a strategy that produces an appropriate n for each value of m her opponent might choose. For instance, to verify (1.1) is true in the natural numbers, Eloise might use the winning strategy: if Abelard picks m, choose $n = m + 1$. If we restrict the choice to only Boolean values, however, Abelard has a winning strategy: he simply picks the value 1. Thus (1.1) is true in the natural numbers, but false if we restrict the choice to Boolean values.

To take an example from calculus, recall that a function f is *continuous* if for every x in its domain, and every $\varepsilon > 0$, there exists a $\delta > 0$ such that for all y,

$$|x - y| < \delta \quad \text{implies} \quad |f(x) - f(y)| < \varepsilon.$$

This definition can be expressed using the quantifier pattern

$$\forall x \forall \varepsilon \exists \delta \forall y (\ldots), \tag{1.2}$$

where the dots stand for an appropriate first-order formula. Using the game-theoretic interpretation, (1.2) is true if for every x and ε chosen by Abelard, Eloise can pick a value for δ such that for every y chosen by Abelard it is the case that . . .

The key feature of game-theoretic semantics is that it relates a central concept of logic (truth) to a central concept of game theory (winning strategy). Once the connection between logic and games has been made, logical principles such as bivalence and the law of excluded middle can be explained using results from game theory. To give one example, the principle of bivalence is an immediate consequence of the Gale-Stewart theorem, which says that in every game of a certain kind there is a player with a winning strategy.

Mathematical logicians have been using game-theoretic semantics implicitly for almost a century. The *Skolem form* of a first-order sentence is obtained by eliminating each existential quantifier, and substituting for the existentially quantified variable a *Skolem term* $f(y_1, \ldots, y_n)$, where f is a fresh function symbol and $y_1, \ldots, y_n$ are the variables upon which the choice of the existentially quantified variable depends. A first-order formula is true in a structure if and only if there are functions satisfying its Skolem form.

For instance the Skolem form of (1.1) is $\forall x \big(x < f(x) \big)$. In the natural numbers, we can take f to be defined by $f(x) = x + 1$, which shows that (1.1) is true. Thus we see that Skolem functions encode Eloise's strategies.

Logic with imperfect information

The game-theoretic perspective allows one to consider extensions of first-order logic that are not obvious otherwise. Independence-friendly logic, the subject of the present volume, is one such extension.

An extensive game with imperfect information is one in which a player may not "see" ("know") all the moves leading up to the current position. Imperfect information is a common phenomenon in card games such as bridge and poker, in which each player knows only the cards on the table and the cards she is holding in her hand.

In order to specify semantic games with imperfect information, the syntax of first-order logic can be extended with slashed sets of variables that indicate which past moves are unknown to the active player. For example, in the independence-friendly sentence

$$\forall x \forall y (\exists z/\{y\}) R(x, y, z), \tag{1.3}$$

the notation $/\{y\}$ indicates that Eloise is not allowed to see the value of y when choosing the value of z.

Imperfect information does not prevent Eloise from performing any particular action she could have taken in the game for the first-order variant of (1.3):

$$\forall x \forall y \exists z R(x, y, z). \tag{1.4}$$

Instead, restricting the information available to the player prevents them from following certain strategies. For instance, in the game for (1.4) played on the natural numbers, Eloise may follow the strategy that takes $z = x + y$. However, this strategy is not available to her in the game for (1.3).

The restriction on Eloise's possible strategies is encoded in the Skolem form of each sentence. For instance, the Skolem form of (1.3) is

$$\forall x \forall y R\big(x, y, f(x)\big),$$

whereas the Skolem form of (1.4) is

$$\forall x \forall y R\big(x, y, f(x, y)\big).$$

The set under the slash in $(\exists z / \{y\})$ indicates that the quantifier is *independent* of the value of y, even though it occurs in the scope of $\forall y$.

Returning to calculus, a function f is *uniformly continuous* if for every x in its domain and every $\varepsilon > 0$, there exists a $\delta > 0$ independent of x such that for all y,

$$|x - y| < \delta \quad \text{implies} \quad |f(x) - f(y)| < \varepsilon.$$

The definition of uniform continuity can be captured by an independence-friendly sentence of the form

$$\forall x \forall \varepsilon \big(\exists \delta / \{x\}\big) \forall y (\ldots),$$

or, equivalently, by a first-order sentence of the form

$$\forall \varepsilon \exists \delta \forall x \forall y (\ldots).$$

Not all independence-friendly sentences are equivalent to a first-order sentence, however. Independence-friendly (IF) logic is related to an earlier attempt to generalize first-order logic made by Henkin [25], who introduced a two-dimensional notation called *branching quantifiers*. For instance, in the branching-quantifier sentence

$$\begin{pmatrix} \forall x \ \exists y \\ \forall z \ \exists w \end{pmatrix} R(x, y, z, w) \tag{1.5}$$

the value of y depends on x, while the value of w depends on z. The Skolem form of the above sentence is given by:

$$\forall x \forall z R\big(x, f(x), z, g(z)\big).$$

We can obtain the same Skolem form from the IF sentence

$$\forall x \exists y \forall z \big(\exists w/\{x,y\}\big) R(x,y,z,w). \tag{1.6}$$

Ehrenfeucht showed that sentences such as (1.5) can define properties that are not expressible in first-order logic [25]. Since branching-quantifier sentences are translatable into IF sentences, IF languages are also more expressive than first-order languages. In fact, IF logic has the same expressive power as existential second-order logic.

The additional expressive power of independence-friendly logic was the main reason why Hintikka advocated its superiority over first-order logic for the foundations of mathematics [28].

Several familiar properties of first-order logic are lost when passing from perfect to imperfect information. They will be discussed in due time. Here we shall briefly consider two such properties. It will be seen that the Gale-Stewart theorem fails for extensive games with imperfect information, and thus there is no guarantee that every IF sentence is either true or false.

One such notorious IF sentence is

$$\forall x \big(\exists y/\{x\}\big) x = y. \tag{1.7}$$

Even on a small domain like the set of Boolean values, Eloise has no way to consistently replicate the choice of Abelard if she is not allowed to see it. Abelard does not have a winning strategy either, though, because Eloise may guess correctly.

Thus, allowing semantic games of imperfect information introduces a third value in addition to true and false. It has been shown that the propositional logic underlying IF logic is precisely Kleene's strong, three-valued logic [31, 34].

Another familiar property of first-order logic that is often taken for granted is that whether an assignment satisfies a formula depends only on the values the assignment gives to the free variables of the formula. In contrast, the meaning of an IF formula can be affected by values assigned to variables that do not occur in the formula at all. This is exemplified by sentences such as

$$\forall x \exists z \big(\exists y/\{x\}\big) x = y. \tag{1.8}$$

In the semantic game for the above sentence, Eloise can circumvent the informational restrictions imposed on the quantifier $(\exists y/\{x\})$ by storing the value of the hidden variable x in the variable z. Thus, the subformula $(\exists y/\{x\})x = y$ has a certain meaning in the context of sentences like (1.7), and a different meaning in the context of sentences like (1.8), where variables other than x may have values.

The failure to properly account for the context-sensitive meanings of IF formulas has resulted in numerous errors appearing in the literature. We shall try to give an accessible and rigorous introduction to the topic.

Traditionally, logicians have been mostly interested in semantic games for which a winning strategy exists. Game theorists, in contrast, have focused more on games for which there is no winning strategy. The most common way to analyze an undetermined game is to allow the players to randomize their strategies, and then calculate the players' expected payoff.

We shall apply the same approach to undetermined IF sentences. While neither player has a winning strategy for the IF sentence (1.7), in a model with exactly two elements, the existential player is as likely to choose the correct element as not, so it seems intuitive to assign the sentence the truth value $1/2$. In a structure with n elements, the probability that the existential player will guess the correct element drops to $1/n$. We will use game-theoretic notions such as mixed strategies and equilibria to provide a solid foundation for such intuitions.

Chapter 2 contains a short primer on game theory that includes all the material necessary to understand the remainder of the book. Chapter 3 presents first-order logic from the game-theoretic perspective. We prove the standard logical equivalences using only the game-theoretic framework, and explore the relationship between semantic games, Skolem functions, and Tarski's classical semantics. Chapter 4 introduces the syntax and semantics of IF logic. Chapter 5 investigates the basic properties of IF logic. We prove independence-friendly analogues to each of the equivalences discussed in Chapter 3, including a prenex normal form theorem. IF logic also shares many of the nice model-theoretic properties of first-order logic. In Chapter 6, we show that IF logic has the same expressive power as existential second-order logic, and the perfect-recall fragment of IF logic has the same expressive power as first-order logic. Chapter 7 analyzes IF formulas whose semantic game is undetermined in terms of mixed strategies and equilibria. In Chapter 8 we discuss the proof that no compositional semantics for IF logic can define its satisfaction relation in terms of single assignments. We also introduce a fragment of IF logic called IF modal logic.

Although it is known that IF logic cannot have a complete deduction system, there have been repeated calls for the development of some kind of proof calculus. The logical equivalences and entailments presented in Chapter 5 form the most comprehensive system to date. They are based on the work of the first author [39, 40], as well as Caicedo, Dechesne, and Janssen [9].

The IF equivalences in Chapter 5 have already proved their usefulness by simplifying the proof of the perfect recall theorem found in Chapter 6, which is due to the third author [52]. The analogue of Burgess' theorem for the perfect-recall fragment of IF logic is due to the first author. The results presented in Chapter 7, due to the third author, generalize results in [52] and extend results in [54].

Acknowledgments

The first author wishes to acknowledge the generous financial support of the Academy of Finland (grant 129208), provided in the context of the European Science Foundation project Logic for Interaction (LINT), which is part of the EUROCORES theme called "LogICCC: Modeling Intelligent Interaction." LINT is a collaborative research project that gathers logicians, computer scientists, and philosophers together in an effort to lay the grounds for a unified account of the logic of interaction.

The first and second authors would like to express their gratitude to the Centre National de la Recherche Scientifique for funding the LINT subproject Dependence and Independence in Logic (DIL) during 2008–2009, and to the Formal Philosophy Seminar at the Institute of History and Philosophy of Science and Technology (Paris 1/CNRS/ENS).

The second author is also grateful for the generous support of the Academy of Finland (grant 1127088).

The authors extend their gratitude to Fausto Barbero, Lauri Hella, Jaakko Hintikka, Antti Kuusisto, and Jonni Virtema for their many helpful comments and suggestions. They also wish to thank everyone who has helped sharpen their thinking about IF logic, including Samson Abramsky, Johan van Benthem, Dietmar Berwanger, Serge Bozon, Julian Bradfield, Xavier Caicedo, Francien Dechesne, Peter van Emde Boas, Thomas Forster, Pietro Galliani, Wilfrid Hodges, Tapani Hyttinen, Theo Janssen, Juha Kontinen, Ondrej Majer, Don Monk, Jan Mycielski,

Anil Nerode, Shahid Rahman, Greg Restall, François Rivenc, Philippe de Rouilhan, B. Sidney Smith, Tero Tulenheimo, Jouko Väänänen, Dag Westerståhl, and Fan Yang.

2
Game theory

According to *A Course in Game Theory*, "a game is a description of strategic interaction that includes the constraints on the actions that the players *can* take and the players' interests, but does not specify the actions that the players *do* take" [45, p. 2]. Classical game theory makes a distinction between *strategic* and *extensive* games. In a strategic game each player moves only once, and all the players move simultaneously. Strategic games model situations in which each player must decide his or her course of action once and for all, without being informed of the decisions of the other players. In an extensive game, the players take turns making their moves one after the other. Hence a player may consider what has already happened during the course of the game when deciding how to move.

We will use both strategic and extensive games in this book, but we consider extensive games first because how to determine whether a first-order sentence is true or false in a given structure can be nicely modeled by an extensive game. It is not necessary to finish the present chapter before proceeding. After reading the section on extensive games, you may skip ahead to Chapter 3. The material on strategic games will not be needed until Chapter 7.

2.1 Extensive games

In an extensive game, the players may or may not be fully aware of the moves made by themselves or their opponents leading up to the current position. When a player knows everything that has happened in the game up till now, we say that he or she has *perfect information*. In the present section we focus on extensive games in which the players always

have perfect information, drawing heavily on the framework found in
Osborne and Rubinstein's classic textbook [45].

2.1.1 Extensive games with perfect information

Definition 2.1   An *extensive game form with perfect information* has
the following components:

- N, a set of *players*.
- H, a set of finite sequences called *histories* or *plays*.
  - If $(a_1, \ldots, a_\ell) \in H$ and $(a_1, \ldots, a_n) \in H$, then for all $\ell < m < n$ we
    must have $(a_1, \ldots, a_m) \in H$. We call $(a_1, \ldots, a_\ell)$ an *initial segment*
    and $(a_1, \ldots, a_n)$ an *extension* of $(a_1, \ldots, a_m)$.
  - A sequence $(a_1, \ldots, a_m) \in H$ is called an *initial history* (or *minimal
    play*) if it has no initial segments in H, and a *terminal history* (or
    maximal play) if it has no extensions in H. We require every history
    to be either terminal or an initial segment of a terminal history. The
    set of terminal histories is denoted Z.
- $P\colon (H - Z) \to N$, the *player function*, which assigns a player $p \in N$
  to each nonterminal history.
  - We imagine that the transition from a nonterminal history $h = (a_1, \ldots, a_m)$ to one of its successors $h^\frown a = (a_1, \ldots, a_m, a)$ in H is
    caused by an *action*. We will identify actions with the final member
    of the successor.
  - The player function indicates whose turn it is to move. For every
    nonterminal history $h = (a_1, \ldots, a_m)$, the player $P(h)$ chooses an
    action a' from the set

$$A(h) = \big\{\, a : (a_1, \ldots, a_m, a) \in H \,\big\},$$

  and play proceeds from $h' = (a_1, \ldots, a_m, a')$.

An *extensive game with perfect information* has the above components,
plus:

- $u_p\colon Z \to \mathbb{R}$, a *utility function* (also called a *payoff function*) for each
  player $p \in N$.   ⊣

Our definition differs from [45, Definition 89.1] in three respects. First,
we do not require initial histories to be empty. Second, we only consider
games that end after a finite number of moves. Third, we use utility
functions to encode the players' preferences rather than working with

preference relations directly. We assume that players always prefer to receive higher payoffs.

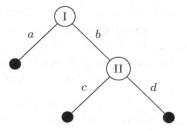

Figure 2.1 An extensive game form with perfect information

When drawing extensive game forms, we label decision points with the active player, and edges with actions. Filled-in nodes represent terminal histories. Figure 2.1 shows the extensive form of a simple two-player game. First, player I chooses between two actions a and b. If she chooses a the game ends. If she chooses b, player II chooses between actions c and d. To obtain an extensive game with perfect information, it suffices to label the terminal nodes with payoffs as shown in Figure 2.2.

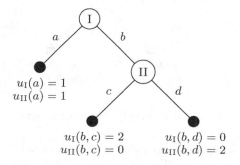

Figure 2.2 An extensive game with perfect information

Notice that the extensive game form depicted in Figure 2.1 has a tree-like structure. A *forest* is a partially ordered set $\mathbb{P} = (P; <)$ such that for all $x \in P$, the set $\{\, y \in P : y < x \,\}$ is well ordered. The *height* of x is just the order type of $\{\, y \in P : y < x \,\}$. A minimal element of a forest has height 0 and is called a *root*; a maximal element is called a *leaf*. The *height* of an entire forest is the least ordinal greater than the height of every element in the forest. A *branch* is a maximal linearly ordered subset of a forest. A forest with a single root is called a *tree*.

For any two histories h and h' of an extensive game form, let $h < h'$ if and only if h is an initial segment of h'. In the game-theoretic literature, it is traditional to draw extensive game forms so that initial histories are at the top, and play proceeds down the branches. An extensive game form has *finite horizon* if the height of its set of histories is finite. All of the games discussed in this book have finite horizon.

Definition 2.2   A two-player extensive game is *strictly competitive* if the players have no incentive to cooperate, that is, if for all $h, h' \in Z$,

$$u_I(h) \geq u_I(h') \quad \text{iff} \quad u_{II}(h') \geq u_{II}(h).$$

A *constant-sum game* is one in which the sum of the players' payoffs is constant, i.e., there exists a $c \in \mathbb{R}$ such that for every terminal history h we have $u_I(h) + u_{II}(h) = c$. When $c = 0$ the game is called *zero sum*.   ⊣

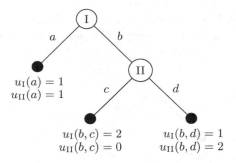

Figure 2.3 A strictly competitive game

In a constant-sum game, any gain for one player is balanced by an offsetting loss for the other. Thus the interests of the players are diametrically opposed. Every constant-sum game is strictly competitive, but not vice versa. For example, the game depicted in Figure 2.3 is strictly competitive, but not constant sum. In a zero-sum game $u_{II}(h) = -u_I(h)$ for every terminal history h.

Definition 2.3   If the only possible payoffs are 1 and 0, we say that player p *wins* a terminal history h if $u_p(h) = 1$, and *loses* if $u_p(h) = 0$. An extensive game is *win-lose* if exactly one player wins each terminal history, in which case we can replace the players' utility functions with

- $u\colon Z \to N$, the *winner function*,

which indicates the winner of each terminal history. When drawing a win-lose game, we label terminal nodes with their winner (Figure 2.4). ⊣

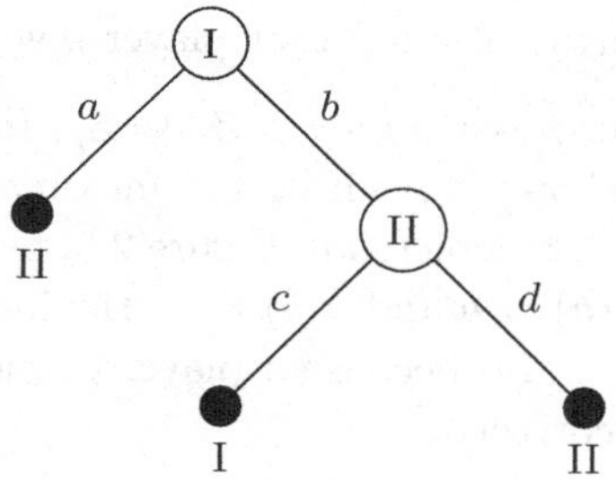

Figure 2.4 A win-lose extensive game

2.1.2 Strategies

When observing many plays of the the same game, we might notice that a given player always makes the same move in a certain position. After observing a few more plays, we may notice other positions in which the player plays consistently. Eventually, we may even be able to predict how the player will move in *any* possible position. At that point, we could say that we know the player's strategy.

Definition 2.4   Let $H_p = P^{-1}(p)$ denote the set of histories where it is player p's turn to move. A *strategy for player p* is a choice function[1]

$$\sigma \in \prod_{h \in H_p} A(h)$$

that tells the player how to move whenever it is his or her turn. A player *follows* a strategy σ during a history $h' = (a_1, \dots, a_n)$ if, whenever $h = (a_1, \dots, a_m) \in H_p$ is an initial segment of h', the history

$$h^\frown \sigma(h) = \big(a_1, \dots, a_m, \sigma(h)\big)$$

is either h' or an initial segment of h'. ⊣

---

[1] Do strategies always exist? Axiom-of-choice skeptics should feel free to substitute "nondeterministic strategy" for "strategy" as necessary. A nondeterministic strategy is a rule that tells a player how to move whenever it is that player's turn — just like a strategy — except that in some situations a nondeterministic strategy might say: "Choose a or b, it doesn't matter."

Three sets of plays are of particular interest:

- H_σ, the plays in which a particular strategy σ is followed;
- $Z_\sigma = H_\sigma \cap Z$, the set of maximal plays in which σ is followed;

and, in a win-lose game,

- $Z_p = u^{-1}(p)$, the maximal plays that player p wins.

A strategy σ for player p is *winning* if $Z_\sigma \subseteq Z_p$. In other words, σ is a winning strategy if and only if p wins every maximal play in which he or she follows σ. In the game depicted in Figure 2.5, player II has a winning strategy defined by $\sigma(a) = b$ and $\sigma(b) = a$. Having a winning strategy is the best possible scenario because a player cannot lose as long as he executes his strategy correctly.

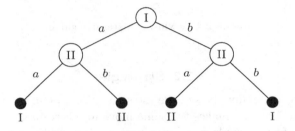

Figure 2.5 Player II has a winning strategy

A win-lose game is *determined* if one of the players has a winning strategy. Determining whether a winning strategy exists for a particular game is not always easy, and can even be independent of the axioms of set theory! Luckily, we will only need to consider a well-behaved class of games for which the answer is already known.

2.1.3 The Gale-Stewart theorem

In their seminal paper, Gale and Stewart [21] used trees to study the class of two-player, win-lose games in which there is a unique initial history, and every terminal history has infinite length ω.

In a Gale-Stewart game it may happen that after finitely many moves one of the players has already won in the sense that he wins every possible continuation of the current history. The other player should obviously try to avoid such histories, and successfully doing so may be enough for her to win. A Gale-Stewart game is *closed* if player I wins every history with the property that all finite initial segments of the history have a terminal extension that she wins. The Gale-Stewart theorem states that every closed game is determined. Since all of the games in this book have finite horizon, we only prove the following special case of the Gale-Stewart theorem.

Theorem 2.5 *Every two-player, win-lose, extensive game with perfect information that has finite horizon and a unique initial history is determined.*

Proof First we extend the winner function to a labeling $\hat{u} \colon H \to \{\text{I}, \text{II}\}$ of every history in a canonical way. Let H_m denote the set of histories of height m. If the game tree has height n then every history in H_{n-1} is terminal. For all $h \in H_{n-1}$ let $\hat{u}(h) = u(h)$. Now suppose $\hat{u}$ has been defined for all $h \in H_m$. Extend $\hat{u}$ to H_{m-1} as follows. If $h \in H_{m-1}$ is terminal, let $\hat{u}(h) = u(h)$. If h is not terminal, define

$$\hat{u}(h) = \begin{cases} P(h) & \text{if there is an } a \in A(h) \text{ such that } \hat{u}(h^\frown a) = P(h), \\ \overline{P(h)} & \text{otherwise,} \end{cases}$$

where $\overline{P(h)}$ denotes the opponent of $P(h)$. When $\hat{u}(h) = p$, we call h a *winning position* for player p.

Let $h_0 \in H_0$ be the unique initial history, and suppose h_0 is a winning position for player p_0. We claim p_0 has a winning strategy σ defined as follows. For all $h \in H_{p_0}$, if h is a winning position for p_0, choose $\sigma(h) \in A(h)$ such that $h^\frown \sigma(h)$ is a winning position for p_0. Otherwise, choose $\sigma(h)$ arbitrarily.

To show σ is a winning strategy we prove that every $h \in H_\sigma$ is a winning position for p_0. The initial history h_0 is a winning position for p_0 by hypothesis. Furthermore, if $h^\frown a \in H_\sigma$, then $h \in H_\sigma$, so by inductive hypothesis h is a winning position for p_0. If $P(h) = p_0$, then $a = \sigma(h)$, and $h^\frown \sigma(h)$ is a winning position for p_0 by construction. If $P(h) = \overline{p_0}$, then $h^\frown a$ must be a winning position for p_0 because otherwise h would have been a winning position for $\overline{p_0}$.

Thus we have shown that every $h \in H_\sigma$ is a winning position for p_0. In particular, every terminal history in which p_0 follows σ is a winning position for p_0. Therefore σ is a winning strategy. $\dashv$

2.1.4 Extensive games with imperfect information

We define extensive games with imperfect information by extending the definition of extensive games with perfect information.

Definition 2.6   An *extensive game form with imperfect information* is a tuple

$$(N, H, P, \{ \sim_p : p \in N \})$$

where N is a set of players, H is a set of histories, P is a player function, and $\sim_p$ is an equivalence relation on H_p with the property that $A(h) = A(h')$ whenever $h \sim_p h'$. When $h \sim_p h'$ we say that h and h' are *indistinguishable for player p*. An *extensive game with imperfect information* is an extensive game form with imperfect information equipped with a utility function $u_p : Z \to \mathbb{R}$ for each player.   ⊣

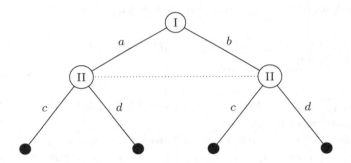

Figure 2.6 An extensive game form with imperfect information

Graphically, we indicate that two histories are indistinguishable for a player by connecting the corresponding decision points with a dotted line. For example, Figure 2.6 depicts an extensive game form with imperfect information in which player II cannot tell whether player I chose a or b on her first move. Therefore player II must always choose c or always choose d, regardless of how his opponent moved. In order to choose an action, however, a player needs to know which actions are possible, hence the requirement that $A(h) = A(h')$ whenever $h \sim_p h'$.

A strategy σ for player p in an extensive game with imperfect information is defined as for an extensive game with perfect information, with the restriction that $\sigma(h) = \sigma(h')$ whenever $h \sim_p h'$. For example, in any extensive game with the form shown in Figure 2.6, player II has only two possible strategies τ and τ' defined by $\tau(a) = c = \tau(b)$ and $\tau'(a) = d = \tau'(b)$, respectively.

Whether player II prefers to follow τ or τ' depends on the payoffs associated with each terminal history. Recall that a winning strategy is one whose owner wins every maximal play in which he or she follows it. Thus, if player II wins both (a, c) and (b, c) then τ is a winning strategy; if he wins both (a, d) and (b, d) then τ' is a winning strategy. Neither τ nor τ' is a winning strategy for the game depicted in Figure 2.7, however. Since player I does not have a winning strategy either, we see that the Gale-Stewart theorem fails for games with imperfect information.

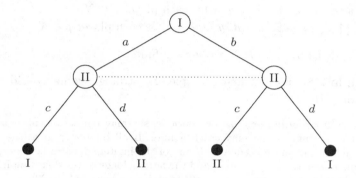

Figure 2.7 An extensive game with imperfect information for which neither player has a winning strategy

2.2 Strategic games

Extensive games emphasize the sequential nature of games, as well as the epistemic states of the players at each decision point. A strategy for an extensive game is a guide that tells its owner how to act in every situation in which he or she is to move. Thus the outcome of an extensive game is determined once the players choose which strategies to follow.

A strategic game abstracts away the internal structure of an extensive game and focuses on the moment the players select their strategies. Once the strategies are fixed, the game ends and the players receive their payoffs.

2.2.1 Pure strategies

In a strategic game, each player selects from a set that includes all possible strategies for that player. Let S_p denote the set of strategies for player p. The elements of S_p are sometimes called pure strategies to distinguish them from mixed strategies, which will be introduced shortly. Pure strategies can be thought of as strategies for some extensive game, or simply as balls in an urn.

Definition 2.7 A *strategic game* has the following components:

- N, a set of players.
- S_p, a set of *(pure) strategies* for each player $p \in N$.
- $u_p \colon \prod_{i \in N} S_i \to \mathbb{R}$, a *utility function* for each player $p \in N$.

A strategic game is *finite* if every S_p is finite. ⊣

Example 2.8 For our first example of a strategic game, consider the game Stag Hunt:

Each of a group of hunters has two options: she may remain attentive to the pursuit of a stag, or she may catch a hare. If all hunters pursue the stag, they catch it and share it equally; if any of the hunters devotes her energy to catching a hare, the stag escapes, and the hare belongs to the defecting hunter alone. Each hunter prefers a share of the stag to a hare. [44, p. 20]

Suppose there are three hunters named I, II, and III. In the scenario described above, the set of strategies for all three hunters is the same, but the payoff each hunter receives depends on the strategies of her fellow hunters, as well as her own strategy. On the one hand, if the third hunter chooses to catch a hare, it is impossible for the first hunter to catch the stag no matter what she or the second hunter does. Thus it is better for the first hunter to catch a hare. On the other hand, if the third hunter pursues the stag, then the first hunter should pursue the stag as well, assuming the second hunter does too (Figure 2.8). ⊣

Player III = stag				Player III = hare		
	stag	hare			stag	hare
stag	2, 2, 2	0, 1, 0		stag	0, 0, 1	0, 1, 1
hare	1, 0, 0	1, 1, 0		hare	1, 0, 1	1, 1, 1

Figure 2.8 The Stag Hunt payoff matrix

	quiet	fink
quiet	−1, −1	−4, 0
fink	0, −4	−3, −3

Figure 2.9 The Prisoner's Dilemma payoff matrix

Since we are mostly interested in applications of strategic games to logic (Chapter 7), we will usually restrict our attention to two-player strategic games. For a two-player strategic game, the adjectives *strictly competitive*, *constant sum*, *zero sum*, and *win-lose* are defined as they are for extensive games, except that histories are replaced by pairs of strategies, e.g., a two-player strategic game is strictly competitive if and only if for all $\sigma, \sigma' \in S_{\mathrm{I}}$ and $\tau, \tau' \in S_{\mathrm{II}}$ we have

$$u_{\mathrm{I}}(\sigma, \tau) \geq u_{\mathrm{I}}(\sigma', \tau') \quad \text{iff} \quad u_{\mathrm{II}}(\sigma', \tau') \geq u_{\mathrm{II}}(\sigma, \tau).$$

Example 2.9 Perhaps the most famous two-player strategic game is the Prisoner's Dilemma:

Two suspects in a major crime are held in separate cells. There is enough evidence to convict each of them of a minor offense, but not enough evidence to convict either of them of the major crime unless one of them acts as an informer against the other (finks). If they both stay quiet, each will be convicted of the minor offense and spend one year in prison. If one and only one of them finks, she will be freed and used as a witness against the other, who will spend four years in prison. If they both fink, each will spend three years in prison. [44, p. 14]

The payoff matrix of the Prisoner's Dilemma is given in Figure 2.9. One can see that the Prisoner's Dilemma is not a strictly competitive game because for both players I and II we have

$$u_p(\text{quiet}, \text{quiet}) > u_p(\text{fink}, \text{fink}).$$

Thus it is in the interest of both suspects to keep quiet. However, if the second prisoner remains quiet, the first prisoner has an incentive to fink because

$$u_{\mathrm{I}}(\text{fink}, \text{quiet}) > u_{\mathrm{I}}(\text{quiet}, \text{quiet}).$$

If the second prisoner finks, then the first prisoner still has an incentive to fink because

$$u_{\mathrm{I}}(\text{fink}, \text{fink}) > u_{\mathrm{I}}(\text{quiet}, \text{fink}).$$

Thus, no matter what the second prisoner does, the first prisoner prefers to rat out her accomplice rather than remain silent. The second prisoner reasons similarly, and so, paradoxically, both prisoners fink on the other, and both are sentenced to three years in prison.   ⊣

Definition 2.10   Let Γ be a two-player strategic game. The pair

$$(\sigma^*, \tau^*) \in S_{\mathrm{I}} \times S_{\mathrm{II}}$$

is an *equilibrium (in pure strategies)* if:

- $u_{\mathrm{I}}(\sigma^*, \tau^*) \geq u_{\mathrm{I}}(\sigma, \tau^*)$ for every pure strategy $\sigma \in S_{\mathrm{I}}$,
- $u_{\mathrm{II}}(\sigma^*, \tau^*) \geq u_{\mathrm{II}}(\sigma^*, \tau)$ for every pure strategy $\tau \in S_{\mathrm{II}}$.   ⊣

There is a visual technique we can use to easily identify the equilibria of a finite, two-player strategic game. In each column, circle the maximum payoffs for player I. For example, in Figure 2.9 we would circle 0 in the first column and -3 in the second column. In each row, circle the maximum payoffs for player II. For example, we would circle 0 in the first row of Figure 2.9 and -3 in the second row. If the maximum payoff occurs more than once in a given column or row, circle every occurrence in that row or column. A pair of pure strategies (σ^*, τ^*) is an equilibrium if and only if both $u_{\mathrm{I}}(\sigma^*, \tau^*)$ and $u_{\mathrm{II}}(\sigma^*, \tau^*)$ are circled. In other words,

$$u_{\mathrm{I}}(\sigma^*, \tau^*) = \max_{\sigma} u_{\mathrm{I}}(\sigma, \tau^*) \quad \text{and} \quad u_{\mathrm{II}}(\sigma^*, \tau^*) = \max_{\tau} u_{\mathrm{II}}(\sigma^*, \tau).$$

If Γ is strictly competitive, the second condition in Definition 2.10 can be rewritten as:

- $u_{\mathrm{I}}(\sigma^*, \tau^*) \leq u_{\mathrm{I}}(\sigma^*, \tau)$ for every pure strategy $\tau \in S_{\mathrm{II}}$.

In this case, (σ^*, τ^*) is an equilibrium if and only if for every $\sigma \in S_{\mathrm{I}}$ and $\tau \in S_{\mathrm{II}}$ we have

$$u_{\mathrm{I}}(\sigma, \tau^*) \leq u_{\mathrm{I}}(\sigma^*, \tau^*) \leq u_{\mathrm{I}}(\sigma^*, \tau). \tag{2.1}$$

Equivalently,

$$\max_{\sigma} u_{\mathrm{I}}(\sigma, \tau^*) = u_{\mathrm{I}}(\sigma^*, \tau^*) = \min_{\tau} u_{\mathrm{I}}(\sigma^*, \tau). \tag{2.2}$$

We can easily generalize Definition 2.10 to games with more than two players. A strategy profile

$$(\sigma_1^*, \ldots, \sigma_n^*) \in \prod_{p \in N} S_p$$

is an equilibrium if and only if no player can improve her utility by unilaterally changing her strategy. By inspecting Figure 2.8 one can see that Stag Hunt has two equilibria. If all the hunters pursue the stag, then each will receive a payoff of 2, which is the maximum possible payoff. If every hunter catches a hare, then each will receive a payoff of 1, which is not optimal, but it is better than nothing. Furthermore, none of the hunters can improve her utility by single-handedly pursuing the stag.

2.2.2 Mixed strategies

Not every strategic game has an equilibrium in pure strategies. In the game Matching Pennies, two players each hold a coin that they secretly turn to heads or tails. Then the coins are revealed simultaneously. If the coins are both heads or both tails, player II pays one dollar to player I; otherwise player I pays one dollar to player II (Figure 2.10).

	heads	tails
heads	1, −1	−1, 1
tails	−1, 1	1, −1

Figure 2.10 The Matching Pennies payoff matrix

Matching Pennies does not have an equilibrium in pure strategies because no matter which strategies the players follow, one of the players will be able to unilaterally improve his or her utility. For example, if both players play heads, player II has an incentive to play tails instead. But if he does, then player I has an incentive to play tails too.

Intuitively, each player should conceal his or her strategy so that the other player cannot take advantage. For example, if the players repeat the game for several rounds, and player I always plays heads, eventually player II will start to play tails. Thus the players should randomize which strategy they select in each round.

Given a set A (called the *sample space*) and a nonempty collection $\mathcal{A} \subseteq \mathscr{P}(A)$ of subsets of A that is closed under complementation, countable unions, and countable intersections, a *probability distribution* over A is a function $\delta \colon \mathcal{A} \to [0,1]$ with the following properties:

(i) $\delta(A) = 1$.
(ii) If $A_1, A_2, \ldots$ are pairwise disjoint sets in $\mathcal{A}$, then

$$\delta \left(\bigcup_{i=1}^{\infty} A_i \right) = \sum_{i=1}^{\infty} \delta(A_i).$$

Elements of the sample space are called *outcomes*, while the subsets in $\mathcal{A}$ are called *events*. If a is an outcome, we write $\delta(a)$ instead of $\delta(\{a\})$. We denote by $\Delta(A)$ the set of all probability distributions over A.

In this book, we will mostly be concerned with finite sample spaces, in which case we may assume that $\mathcal{A} = \mathscr{P}(A)$, i.e., every set of outcomes is an event. Let δ be a probability distribution over a finite sample space. The *support* of δ is the set of outcomes to which δ assigns nonzero probability. We say δ is *uniform* if it assigns equal probability to all the outcomes in its support.

Definition 2.11 Let $\Gamma = \big(N, \{S_p : p \in N\}, \{u_p : p \in N\}\big)$ be a strategic game. A *mixed strategy for player $p \in N$* is a probability distribution $\mu \in \Delta(S_p)$.   ⊣

Let Γ be a two-player strategic game, and let μ and ν be mixed strategies for players I and II, respectively. The probability that player I follows $\sigma \in S_I$ is precisely $\mu(\sigma)$, and the probability that player II follows $\tau \in S_{II}$ is $\nu(\tau)$. We assume that the players choose their strategies independently of each other. Thus the probability that player I follows σ and player II follows τ is the product $\mu(\sigma)\nu(\tau)$.

After many rounds in which player I employs μ and player II employs ν, the average payoff a player can expect to receive is the weighted average of the payoffs he or she receives when the players follow a given pair of pure strategies. For example, when S_I and S_{II} are finite, the *expected utility function* for player p is

$$U_p(\mu, \nu) = \sum_{\sigma \in S_I} \sum_{\tau \in S_{II}} \mu(\sigma)\nu(\tau)u_p(\sigma, \tau).$$

If Γ is a finite c-sum game, that is, $u_I(\sigma, \tau) + u_{II}(\sigma, \tau) = c$ for all $\sigma \in S_I$ and $\tau \in S_{II}$, one can easily calculate that

$$U_I(\mu, \nu) + U_{II}(\mu, \nu) = c.$$

Example 2.12 In the game Matching Pennies, let μ be the mixed strategy for player I defined by $\mu(\text{heads}) = p$ and $\mu(\text{tails}) = 1 - p$. Let ν be the mixed strategy for player II defined by $\nu(\text{heads}) = q$ and $\nu(\text{tails}) = 1 - q$. The expected utility for player I is:

$$\begin{aligned}
U_I(\mu, \nu) &= \sum_{\sigma \in S_I} \sum_{\tau \in S_{II}} \mu(\sigma)\nu(\tau)u_I(\sigma, \tau) \\
&= pq - p(1 - q) - (1 - p)q + (1 - p)(1 - q) \\
&= (2p - 1)(2q - 1).
\end{aligned}$$

For example, if $p = 2/3$ and $q = 1/4$, then $U_{\mathrm{I}}(\mu, \nu) = -1/6$. Since Matching Pennies is a zero-sum game it follows that $U_{\mathrm{II}}(\mu, \nu) = 1/6$. ⊣

We can simulate a pure strategy σ with a mixed strategy μ that assigns σ probability 1. We will commonly identify such "degenerate" mixed strategies with the unique pure strategy in their support. If $\sigma \in S_{\mathrm{I}}$ and $\nu \in \Delta(S_{\mathrm{II}})$,

$$U_p(\sigma, \nu) = \sum_{\tau \in S_{\mathrm{II}}} \nu(\tau) u_p(\sigma, \tau).$$

Similarly, if $\mu \in \Delta(S_{\mathrm{I}})$ and $\tau \in S_{\mathrm{II}}$,

$$U_p(\mu, \tau) = \sum_{\sigma \in S_{\mathrm{I}}} \mu(\sigma) u_p(\sigma, \tau).$$

It follows immediately from the definitions that

$$U_p(\mu, \nu) = \sum_{\sigma \in S_{\mathrm{I}}} \mu(\sigma) U_p(\sigma, \nu) = \sum_{\tau \in S_{\mathrm{II}}} \nu(\tau) U_p(\mu, \tau). \tag{2.3}$$

With these definitions in place, we can now say what it means for a pair of mixed strategies to be an equilibrium.

Definition 2.13 Let Γ be a two-player strategic game with $\mu^* \in \Delta(S_{\mathrm{I}})$ and $\nu^* \in \Delta(S_{\mathrm{II}})$. The pair (μ^*, ν^*) is an *equilibrium (in mixed strategies)* if it satisfies the following conditions:

- $U_{\mathrm{I}}(\mu^*, \nu^*) \geq U_{\mathrm{I}}(\mu, \nu^*)$ for every mixed strategy $\mu \in \Delta(S_{\mathrm{I}})$,
- $U_{\mathrm{II}}(\mu^*, \nu^*) \geq U_{\mathrm{II}}(\mu^*, \nu)$ for every mixed strategy $\nu \in \Delta(S_{\mathrm{II}})$. ⊣

If Γ is strictly competitive, then (μ^*, ν^*) is an equilibrium if and only if for all $\mu \in \Delta(S_{\mathrm{I}})$ and $\nu \in \Delta(S_{\mathrm{II}})$,

$$U_{\mathrm{I}}(\mu, \nu^*) \leq U_{\mathrm{I}}(\mu^*, \nu^*) \leq U_{\mathrm{I}}(\mu^*, \nu). \tag{2.4}$$

Equivalently,

$$\max_{\mu} U_{\mathrm{I}}(\mu, \nu^*) = U_{\mathrm{I}}(\mu^*, \nu^*) = \min_{\nu} U_{\mathrm{I}}(\mu^*, \nu). \tag{2.5}$$

Example 2.14 The pair (μ, ν) of mixed strategies in Example 2.12 according to which player I plays heads with probability 2/3 and player II plays heads with probability 1/4 is not an equilibrium. Keeping μ fixed, player II can improve his expected utility to 1/3 by always playing tails ($q = 0$). However, if μ^* is the mixed strategy according to which player I plays heads with probability $p = 1/2$, then

$$U_{\mathrm{I}}(\mu^*, \nu) = 0 = U_{\mathrm{II}}(\mu^*, \nu)$$

regardless of q. Similarly, if ν^* is the mixed strategy according to which player II plays heads with probability $q = 1/2$, then

$$U_I(\mu, \nu^*) = 0 = U_{II}(\mu, \nu^*)$$

regardless of p. Thus (μ^*, ν^*) is an equilibrium.

In fact (μ^*, ν^*) is the only equilibrium for Matching Pennies. If player I plays heads with probability $p > 1/2$, then player II can maximize his utility by always playing tails. If $p < 1/2$, then player II maximizes his utility by always playing heads. Similarly, if $q > 1/2$ then player I should always play heads, and if $q < 1/2$ she should always play tails.   ⊣

Every equilibrium in pure strategies is also an equilibrium in mixed strategies. Thus Stag Hunt shows that a strategic game can have multiple mixed-strategy equilibria, and that a given player may receive different payoffs from different equilibria. A strictly competitive game can also have multiple equilibria, but the players always receive the same payoffs.

Proposition 2.15 *If (μ, ν) and (μ', ν') are two equilibria in mixed strategies for a strictly competitive strategic game, then*

$$U_p(\mu, \nu) = U_p(\mu', \nu').$$

Proof If (μ, ν) and (μ', ν') are both equilibria, then

$$U_I(\mu, \nu) \leq U_I(\mu, \nu') \leq U_I(\mu', \nu') \leq U_I(\mu', \nu) \leq U_I(\mu, \nu).$$

Hence $U_I(\mu, \nu) = U_I(\mu', \nu')$. Likewise, $U_{II}(\mu, \nu) = U_{II}(\mu', \nu')$.   ⊣

Note that Proposition 2.15 does not guarantee that a strictly competitive game has an equilibrium, only that every equilibrium results in the same payoffs, if any such equilibria exist. Luckily, von Neumann [43] proved that for a special class of games, equilibria always exist. Known as the minimax theorem, this result is considered by many to be the first major theorem in game theory.

Theorem 2.16 (Minimax) *Every finite, two-person, zero-sum game has an equilibrium in mixed strategies.*   ⊣

Nash later proved that every finite strategic game has a mixed-strategy equilibrium [42]. The notion of equilibrium has been associated with Nash's name ever since. The theory developed in this volume, however, depends only on the minimax theorem.

The following proposition allows us to extend the minimax theorem to constant-sum games.

Proposition 2.17 *Let Γ and Γ' be finite, two-player strategic games involving the same players and the same strategy sets. Let $f(x) = ax + b$ be an affine transformation, where $a, b \in \mathbb{R}$ and $a > 0$. If for all $\sigma \in S_\mathrm{I}$ and $\tau \in S_\mathrm{II}$, $u'_p(\sigma, \tau) = f\big(u_p(\sigma, \tau)\big)$, where u_p is the utility function of player p in Γ, and u'_p is the utility function of player p in Γ', then every mixed-strategy equilibrium in Γ is a mixed-strategy equilibrium in Γ'.*

Proof We write U'_p for the expected utility of player p in Γ'. It is easy to show that for every $\mu \in \Delta(S_\mathrm{I})$ and $\nu \in \Delta(S_\mathrm{II})$,

$$U'_p(\mu, \nu) = aU_p(\mu, \nu) + b.$$

Let (μ^*, ν^*) be a mixed-strategy equilibrium in Γ. Then for every $\mu \in \Delta(S_\mathrm{I})$ we have

$$U'_\mathrm{I}(\mu^*, \nu^*) = aU_\mathrm{I}(\mu^*, \nu^*) + b \geq aU_\mathrm{I}(\mu, \nu^*) + b = U'_\mathrm{I}(\mu, \nu^*).$$

Similarly, for every $\nu \in \Delta(S_\mathrm{II})$ we have $U'_\mathrm{II}(\mu^*, \nu^*) \geq U'_\mathrm{II}(\mu^*, \nu)$. Thus (μ^*, ν^*) is an equilibrium in Γ'.   ⊣

2.2.3 Identifying equilibria

When presented with a pair (μ^*, ν^*) of mixed strategies for a strategic game, determining whether it is an equilibrium seems a daunting task because one must compare it to all other pairs of mixed strategies for the same game. The next proposition shows that it suffices to compare (μ^*, ν^*) against pairs of the form (σ, ν^*) and (μ^*, τ), where σ and τ are pure strategies [44, p. 116].

Proposition 2.18 *In a finite, two-player strategic game, the pair $(\mu^*, \nu^*) \in \Delta(S_\mathrm{I}) \times \Delta(S_\mathrm{II})$ is a mixed-strategy equilibrium if and only if all of the following conditions hold:*

(1) $U_\mathrm{I}(\mu^*, \nu^*) = U_\mathrm{I}(\sigma, \nu^*)$ for every $\sigma \in S_\mathrm{I}$ in the support of μ^*,
(2) $U_\mathrm{II}(\mu^*, \nu^*) = U_\mathrm{II}(\mu^*, \tau)$ for every $\tau \in S_\mathrm{II}$ in the support of ν^*,
(3) $U_\mathrm{I}(\mu^*, \nu^*) \geq U_\mathrm{I}(\sigma, \nu^*)$ for every $\sigma \in S_\mathrm{I}$ outside the support of μ^*,
(4) $U_\mathrm{II}(\mu^*, \nu^*) \geq U_\mathrm{II}(\mu^*, \tau)$ for every $\tau \in S_\mathrm{II}$ outside the support of ν^*.

Proof Suppose (μ^*, ν^*) is an equilibrium in mixed strategies.

(1) Let us consider only the strategies in the support of μ_I,

$$S_\mathrm{I}^* = \big\{\, \sigma \in S_\mathrm{I} : \mu^*(\sigma) > 0 \,\big\}.$$

For every $\sigma \in S_I$ we have $U_I(\sigma, \nu^*) \leq U_I(\mu^*, \nu^*)$ because (μ^*, ν^*) is an equilibrium. By Equation (2.3) we know that

$$\sum_{\sigma \in S_I^*} \mu^*(\sigma) U(\sigma, \nu^*) = U_I(\mu^*, \nu^*).$$

If $U_I(\sigma, \nu^*) < U_I(\mu^*, \nu^*)$ for every $\sigma \in S_I^*$ we would have

$$\sum_{\sigma \in S_I^*} \mu^*(\sigma) U(\sigma, \nu^*) < U_I(\mu^*, \nu^*),$$

which is not the case. Hence there is a pure strategy $\sigma^* \in S_I^*$ such that $U_I(\sigma^*, \nu^*) = U_I(\mu^*, \nu^*)$. Now suppose for the sake of a contradiction that for some $\sigma' \in S_I^*$ we have $U_I(\sigma', \nu^*) < U_I(\sigma^*, \nu^*)$. Then

$$U_I(\mu^*, \nu^*) = \sum_{\sigma \in S_I^*} \mu^*(\sigma) U_I(\sigma, \nu^*)$$

$$= \mu^*(\sigma') U_I(\sigma', \nu^*) + \sum_{\sigma \in S_I^* - \{\sigma'\}} \mu^*(\sigma) U_I(\sigma, \nu^*)$$

$$< \mu^*(\sigma') U_I(\sigma^*, \nu^*) + \sum_{\sigma \in S_I^* - \{\sigma'\}} \mu^*(\sigma) U_I(\sigma, \nu^*).$$

That is, player I improves her expected utility by playing σ^* instead of σ', contradicting the fact that (μ^*, ν^*) is an equilibrium. Thus $U_I(\sigma, \nu^*) = U(\mu^*, \nu^*)$ for every $\sigma \in S_I^*$.

The proof of (2) is similar, while (3) and (4) are immediate.

For the converse, suppose (μ^*, ν^*) satisfies conditions (1)–(4). The sets

$$S_I^* = \big\{ \sigma \in S_I : \mu^*(\sigma) > 0 \big\} \quad \text{and} \quad \overline{S_I} = \big\{ \sigma \in S_I : \mu^*(\sigma) = 0 \big\}$$

are disjoint and $S_I^* \cup \overline{S_I} = S_I$, so for any $\mu \in \Delta(S_I)$,

$$U_I(\mu, \nu^*) = \sum_{\sigma \in S_I^*} \mu(\sigma) U_I(\sigma, \nu^*) + \sum_{\sigma \in \overline{S_I}} \mu(\sigma) U_I(\sigma, \nu^*)$$

$$\leq \sum_{\sigma \in S_I^*} \mu(\sigma) U_I(\mu^*, \nu^*) + \sum_{\sigma \in \overline{S_I}} \mu(\sigma) U_I(\mu^*, \nu^*)$$

$$= U_I(\mu^*, \nu^*) \sum_{\sigma \in S_I} \mu(\sigma)$$

$$= U_I(\mu^*, \nu^*)$$

by conditions (1) and (3). A similar argument shows that $U_{II}(\mu^*, \nu) \leq U_{II}(\mu^*, \nu^*)$ for every $\nu \in \Delta(S_{II})$.   ⊣

	τ_1	τ_2	τ_3	τ_4
σ_1	1	0	1	0
σ_2	1	1	0	0
σ_3	0	1	1	0
σ_4	0	0	0	1

Figure 2.11 Player I's payoff matrix

Example 2.19 Let Γ be a two-player, win-lose game for which player I's utility function is depicted in Figure 2.11. Consider the pair of mixed strategies (μ^*, ν^*), where $\mu^* \in \Delta(S_\mathrm{I})$ is defined by

$$\mu^*(\sigma_i) = \begin{cases} 1/5 & \text{if } \sigma_i \in \{\sigma_1, \sigma_2, \sigma_3\}, \\ 2/5 & \text{if } \sigma_i \in \{\sigma_4\}, \end{cases}$$

and $\nu^* \in \Delta(S_\mathrm{II})$ is defined by

$$\nu^*(\tau_j) = \begin{cases} 1/5 & \text{if } \tau_j \in \{\tau_1, \tau_2, \tau_3\}, \\ 2/5 & \text{if } \tau_j \in \{\tau_4\}. \end{cases}$$

We leave it as an exercise to compute that $U_\mathrm{I}(\mu^*, \nu^*) = 2/5$. To see that (μ^*, ν^*) is indeed an equilibrium, consider a strategy $\sigma_i \in S_\mathrm{I}$. If $i = 1$, then

$$U_\mathrm{I}(\sigma_1, \nu^*) = \sum_{j=1}^{4} \nu^*(\tau_j) u_\mathrm{I}(\sigma_1, \tau_j) = \nu^*(\tau_1) + \nu^*(\tau_3) = \frac{2}{5}.$$

The calculations for $i \in \{2, 3\}$ are similar. If $i = 4$, then we have

$$U_\mathrm{I}(\sigma_4, \nu^*) = \nu^*(\tau_4) = \frac{2}{5}.$$

As an exercise, the reader should check that $U_\mathrm{II}(\mu^*, \tau_j) = 3/5$ for all $1 \le j \le 4$. (Note that the payoff matrix for player II is the inverse of the one shown in Figure 2.11.) Thus (μ^*, ν^*) is an equilibrium. ⊣

3
First-order logic

We assume the reader is familiar with first-order logic. Nevertheless, we feel it worthwhile to collect the main definitions and results in the present chapter so that we may refer to them throughout the remainder of the book. Doing so will serve the dual purpose of making the book self-contained and preparing the reader for what lies ahead.

3.1 Syntax

Let $\{x_0, x_1, x_2, \ldots\}$ be a countably infinite set of variables. We will use $x, y, z, \ldots$ as meta-variables ranging over the variables in this set. The symbols $U, V, W, \ldots$ range over finite sets of variables.

Definition 3.1 A *vocabulary* is a set of relation symbols and function symbols. Each symbol is associated with a natural number, called its *arity*, that indicates the number of arguments the symbol accepts. Nullary function symbols are called *constant symbols*, and will often be treated separately. Unary relation symbols are sometimes called *predicates*. ⊣

The function symbols in a vocabulary can be combined with variables to form more complicated expressions called terms.

Definition 3.2 Let L be a vocabulary. The set of L-*terms* is generated by the finite application of the following rules:

- Every variable is an L-term.
- Every constant symbol in L is an L-term.
- If f is an n-ary function symbol in L and $t_1, \ldots, t_n$ are L-terms, then $f(t_1, \ldots, t_n)$ is an L-term. ⊣

Relation symbols can be combined with terms to form (atomic) formulas. Formulas can be combined with logical connectives and quantifiers to form (compound) formulas.

Definition 3.3 The first-order language generated by the vocabulary L, denoted FO_L, is generated by the finite application of the following rules:

- If t_1 and t_2 are L-terms, then $(t_1 = t_2) \in \mathrm{FO}_L$.
- If R is an n-ary relation symbol in L and $t_1, \ldots, t_n$ are L-terms, then $R(t_1, \ldots, t_n) \in \mathrm{FO}_L$.
- If $\varphi \in \mathrm{FO}_L$, then $\neg\varphi \in \mathrm{FO}_L$.
- If $\varphi, \varphi' \in \mathrm{FO}_L$, then $(\varphi \vee \varphi') \in \mathrm{FO}_L$ and $(\varphi \wedge \varphi') \in \mathrm{FO}_L$.
- If $\varphi \in \mathrm{FO}_L$ and x is a variable, then $\exists x\varphi \in \mathrm{FO}_L$ and $\forall x\varphi \in \mathrm{FO}_L$. ⊣

The elements of FO_L are called FO_L *formulas*. A *first-order formula* is an FO_L formula for some vocabulary L. When the vocabulary is irrelevant or clear from context we will not mention it explicitly. Formulas of the form $(t_1 = t_2)$ or $R(t_1, \ldots, t_n)$ are called *atomic*. It will sometimes be convenient to let the symbol $\circ$ range over the set of connectives $\{\vee, \wedge\}$, and let Q range over the set of quantifiers $\{\exists, \forall\}$. By $\overline{Q}$ we mean the dual of Q, which is to say $\overline{\exists} = \forall$ and $\overline{\forall} = \exists$. When writing first-order formulas we will be flexible in our use of brackets.

Definition 3.4 Let φ be a first-order formula. The set of *subformulas* of φ, denoted $\mathrm{Subf}(\varphi)$, is defined recursively:

$$\mathrm{Subf}(\psi) = \{\psi\} \quad (\psi \text{ atomic}),$$
$$\mathrm{Subf}(\neg\psi) = \{\neg\psi\} \cup \mathrm{Subf}(\psi),$$
$$\mathrm{Subf}(\psi \circ \psi') = \{\psi \circ \psi'\} \cup \mathrm{Subf}(\psi) \cup \mathrm{Subf}(\psi'),$$
$$\mathrm{Subf}(Qx\psi) = \{Qx\psi\} \cup \mathrm{Subf}(\psi).$$

The set of atomic subformulas of φ is denoted $\mathrm{Atom}(\varphi)$. The set of existentially quantified subformulas is denoted $\mathrm{Subf}_\exists(\varphi)$, while the set of universally quantified subformulas is denoted $\mathrm{Subf}_\forall(\varphi)$. ⊣

We will treat each instance of a particular subformula as being distinct.[1] For example, in the formula $\psi \vee \psi$ we distinguish between the left and right disjunct. If ψ is a proper subformula of $Qx\varphi$ we say ψ lies within the *scope* of Qx, as does every variable and quantifier that occurs

---

[1] There are several ways to enforce this property, one of which would, for instance, rely on an indexing of the brackets enclosing subformulas: $(\psi)_i$.

in ψ. If $Q'y$ lies within the scope of Qx we say that $Q'y$ is *subordinate* to Qx and that Qx is *superordinate* to $Q'y$.

Definition 3.5   A particular occurrence of a variable x is *free* in φ if it does not lie within the scope of any quantifier of the form Qx. If φ is atomic, all its variables are free. For compound formulas:

$$\text{Free}(\neg\varphi) = \text{Free}(\varphi),$$
$$\text{Free}(\varphi \circ \psi) = \text{Free}(\varphi) \cup \text{Free}(\psi),$$
$$\text{Free}(Qx\varphi) = \text{Free}(\varphi) - \{x\}. \qquad\dashv$$

An occurrence of a variable is *bound* if it is not free. Specifically, an occurrence of a variable is bound by the innermost quantifier in whose scope it lies. That is, an occurrence of x in the formula $Qx\varphi$ is bound by Qx if and only if $x \in \text{Free}(\varphi)$. The set of bound variables of φ is denoted $\text{Bound}(\varphi)$. For example, in the formula

$$\forall x \exists y (x + y \le z),$$

the variables x and y are bound, whilst z is free. In the formula

$$\text{Red}(x) \vee \exists x\, \text{Green}(x)$$

the variable x occurs both free and bound. A formula with no free variables is called a *sentence*.

3.2 Models

It is now time to give content to the relation symbols and function symbols that constitute the vocabulary of our first-order languages. We do this by introducing mathematical structures, or models. A model is a nonempty set equipped with relations and operations that interpret the relation symbols and functions symbols of a vocabulary.

Definition 3.6   Let $L = \{R, \ldots, f, \ldots, c, \ldots\}$ be a vocabulary. An *L-structure* is an object

$$\mathbb{M} = (M; R^{\mathrm{M}}, \ldots, f^{\mathrm{M}}, \ldots, c^{\mathrm{M}}, \ldots)$$

where M is a nonempty set called the *universe* of $\mathbb{M}$. The *size* of a structure refers to the cardinality of its universe. If R is an n-ary relation symbol, then R^{M} is an n-ary relation on M called the *interpretation of* R, and if f is an n-ary function symbol, then $f^{\mathrm{M}} \colon M^n \to M$ is an

n-ary function on M called the *interpretation of f*. If c is a constant symbol, then $c^M : M^0 \to M$ is a constant function. We will usually identify c^M with its unique value. We will use the terms *structure* and *model* interchangeably.   ⊣

Let φ be an FO_L formula, let M be an L-structure, and M′ an L'-structure. If $L \subseteq L'$ we say that M′ is *suitable* for φ because it has interpretations for all the relation and function symbols that occur in φ. Furthermore, if M and M′ have the same universe, and for every relation symbol R in L we have $R^M = R^{M'}$, and for every function symbol f in L we have $f^M = f^{M'}$, then M′ is an *expansion* of M to L', and M is the *reduct* of M′ to L (written M = M′↾L).

A model tells us how to interpret the symbols in a vocabulary, but it does not tell us the values of any variables. When pondering the atomic formula $R(x, y, z)$ it helps to know which individuals the variables x, y and z refer to.

Definition 3.7 Let M be a structure. An *assignment in* M is a partial function from the set of variables $\{x_0, x_1, x_2, \ldots\}$ to M. If s is an assignment in M, and $a \in M$, then $s(x_i/a)$ denotes the assignment with domain $\mathrm{dom}(s) \cup \{x_i\}$ defined by:

$$s(x_i/a)(x_j) = \begin{cases} s(x_j) & \text{if } i \neq j, \\ a & \text{if } i = j. \end{cases}$$

That is, $s' = s(x/a)$ is exactly like s except $s'(x) = a$. Notice that if $x \in \mathrm{dom}(s)$ the value that s assigns to x is overwritten when we assign it a new value.

The notation $s(x, y, z) = (a, b, c)$ is an abbreviation for $s(x) = a$, $s(y) = b$, and $s(z) = c$. We shall every now and then extend an assignment to a mapping from terms to individuals:

$$s(c) = c^M,$$
$$s\big(f(t_1, \ldots, t_n)\big) = f^M\big(s(t_1), \ldots, s(t_n)\big).$$ ⊣

We now have enough pieces of the puzzle to determine whether an atomic formula is true or false when its variables are taken to refer to specific individuals.

Definition 3.8 Let L be a vocabulary, $\mathbb{M}$ an L-structure, and s an assignment in $\mathbb{M}$. Then

$$\mathbb{M}, s \models (t_1 = t_2) \quad \text{iff} \quad s(t_1) = s(t_2),$$
$$\mathbb{M}, s \models R(t_1, \ldots, t_n) \quad \text{iff} \quad \big(s(t_1), \ldots, s(t_n)\big) \in R^{\mathbb{M}}. \qquad \dashv$$

For example, let $\mathbb{N} = \{\, \omega; \, +^{\mathbb{N}}, \, \cdot^{\mathbb{N}}, \, \leq^{\mathbb{N}} \,\}$ be the set of natural numbers equipped with the standard operations of addition and multiplication along with the normal ordering. Let $s(x, y) = (3, 7)$. Then

$$\mathbb{N}, s \models x \leq y, \qquad\qquad \mathbb{N}, s \not\models y \leq x,$$
$$\mathbb{N}, s \models x + y = y + x, \qquad \mathbb{N}, s \not\models (x \cdot y) + x = (x + y) \cdot x$$

Before continuing, the reader should practice constructing models and testing whether a given model and assignment satisfy various atomic formulas.

3.3 Game-theoretic semantics

In *Philosophical Investigations* [68], Wittgenstein uses games to explain how we use and learn language. In a nutshell, words and sentences acquire their meaning from their use in various language-games. Later Hintikka applied Wittgenstein's analysis to first-order logic. He proposed game-theoretic semantics [26, 27] as an alternative to the recursive definition of truth *via* satisfaction advocated by Tarski [58, 59, 61]. In this section we use extensive games with perfect information to formalize the game-theoretic semantics for first-order logic.

Consider the following dialogue:[2]

ABELARD	Eloise, tell me, is there a smallest natural number?
ELOISE	Yes.
ABELARD	Which is it?
ELOISE	Zero.
ABELARD	Correct. Any other number I might choose would be greater. Is there a greatest natural number?
ELOISE	No.
ABELARD	Why not?
ELOISE	Because no matter which number I pick, you can always find a greater one.

---

[2] Peter Abelard was a renowned medieval logician; Eloise was his student and lover [1].

In the first part of the dialogue, Eloise asserts the truth of the sentence

$$\exists x \forall y (x \leq y)$$

when the quantifiers range over the natural numbers and the symbol $\leq$ is interpreted by the normal ordering. When asked to justify her claim, she responds by picking a natural number. Eloise must choose carefully, of course. If she had picked any number besides zero, Abelard could have found a smaller one. In the second part of the dialogue, Eloise denies the sentence

$$\exists x \forall y (y \leq x).$$

When asked to give her reason for denying it, she explains that her teacher could always win a game in which they each pick a natural number, with Eloise choosing first, and the player who chooses the greater number wins.

As early as 1898, Peirce noticed that quantifiers can be interpreted as moves in a game. In his second Cambridge Conferences lecture, titled "Types of Reasoning," he remarks:

When I say "every man dies," I say you may pick out your man for yourself and provided he belongs to *this here* world you will find he will die. The "some" supposes a selection from "this here" world to be made by the *deliverer* of the proposition, or made in his interest. The "every" *transfers* the function of selection to the *interpreter* of the proposition, or to anybody acting in his interest. [46, pp. 129–30]

Calculus instructors continually rediscover the connection between quantifiers and games. One professor of our acquaintance challenges his students by saying: "The Devil chooses an $\varepsilon > 0$! Find a $\delta > 0$ such that..."

Logical connectives can also be considered as moves in a game. If Eloise asserts a disjunction she may choose which disjunct she wishes to verify. If she asserts a conjunction she must verify the conjunct of Abelard's choosing.

By now the reader's intuition should be robust enough to play the semantic game for any first-order formula. Nevertheless, it is important to give a formal definition. For the moment, we will only consider formulas in which the negation symbol $\neg$ does not appear.

Definition 3.9 Let φ be a negation-free formula, $\mathbb{M}$ a suitable structure and s an assignment in $\mathbb{M}$ whose domain contains $\mathrm{Free}(\varphi)$. The *semantic game* $G(\mathbb{M}, s, \varphi)$ is a win-lose extensive game with perfect information:

- There are two players, Eloise ($\exists$) and Abelard ($\forall$).
- The set of histories is $H = \bigcup\{\, H_\psi : \psi \in \mathrm{Subf}(\varphi)\,\}$, where H_ψ is defined recursively:
  - $H_\varphi = \{(s, \varphi)\}$,
  - if ψ is $\chi_1 \circ \chi_2$, then $H_{\chi_i} = \{\, h^\frown \chi_i : h \in H_{\chi_1 \circ \chi_2}\,\}$,
  - if ψ is $Qx\chi$, then $H_\chi = \{\, h^\frown(x,a) : h \in H_{Qx\chi},\, a \in M\,\}$.

Observe that (s, φ) is the unique initial history. The assignment s is called the *initial assignment*. Every history h' induces an assignment $s_{h'}$ extending or modifying the initial assignment:

$$
s_{h'} = \begin{cases} s & \text{if } h' = (s, \varphi), \\ s_h & \text{if } h' = h^\frown \chi \text{ for some } \chi \in \mathrm{Subf}(\varphi), \\ s_h(x/a) & \text{if } h' = h^\frown(x,a) \text{ for some } a \in M. \end{cases}
$$

- Once a play has reached an atomic formula the game ends:
$$
Z = \bigcup\{\, H_\chi : \chi \in \mathrm{Atom}(\varphi)\,\}.
$$

- Disjunctions and existential quantifiers are decision points for Eloise, while conjunctions and universal quantifiers are decision points for Abelard:
$$
P(h) = \begin{cases} \exists & \text{if } h \in H_{\chi \vee \chi'} \text{ or } h \in H_{\exists x \chi}, \\ \forall & \text{if } h \in H_{\chi \wedge \chi'} \text{ or } h \in H_{\forall x \chi}. \end{cases}
$$

- Eloise wins a maximal play $h \in H_\chi$ if the atomic formula χ is satisfied by the current assignment; Abelard wins if it is not:
$$
u(h) = \begin{cases} \exists & \text{if } \mathrm{M}, s_h \models \chi, \\ \forall & \text{if } \mathrm{M}, s_h \not\models \chi. \end{cases} \qquad \dashv
$$

Consider the semantic game for $\exists x \forall y (x \leq y)$ interpreted in $\mathbb{N}$. For convenience, let φ denote the original sentence, and let ψ denote the subformula $\forall y(x \leq y)$. Assume the initial assignment is empty, so that $H_\varphi = \{(\varnothing, \varphi)\}$. Eloise moves first, choosing a value for x. Thus

$$
H_\psi = \{\, (\varnothing, \varphi, (x,a)) : a \in \omega\,\}.
$$

Then Abelard picks a value for y, and the game ends:

$$
Z = \{\, (\varnothing, \varphi, (x,a), (y,b)) : a,b \in \omega\,\}.
$$

Eloise wins if $a \leq^{\mathbb{N}} b$; otherwise Abelard wins. We see immediately that Eloise has a winning strategy $\sigma(\varnothing, \varphi) = (x, 0)$ (see Figure 3.1).[3]

[3] Note that we omitted the outer parentheses when applying the strategy σ to the

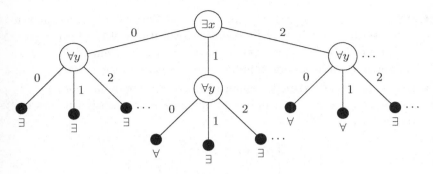

Figure 3.1 The semantic game for $\exists x \forall y (x \leq y)$ in $\mathbb{N}$

Now consider the semantic game for $\exists x \forall y (y \leq x)$. The collection of histories is the same as before, but this time Eloise wins if $b \leq^{\mathbb{N}} a$. Unfortunately for her, Abelard has a winning strategy $\tau(\varnothing, \varphi, (x, a)) = (y, a + 1)$.

3.3.1 Negation

For clarity of presentation, we delayed discussion of the game rules for negated formulas. When Eloise asserts $\neg\varphi$ she is denying φ. In other words, she is claiming that, were Abelard to assert φ, she would be able to refute him. Thus a negation symbol indicates the role-reversal of the players. Let φ be a first-order formula involving negations. While Eloise's and Abelard's identities remain fixed during the semantic game $G(\mathbb{M}, s, \varphi)$, their roles do not. In a given position Eloise may be trying to verify or falsify the current subformula, with Abelard always attempting to thwart her. Let us call the player trying to verify the current formula the *verifier*, and his or her opponent the *falsifier*. In order to generalize Definition 3.9 to all first-order formulas, we need to modify the player function and the winner function to account for possible role-reversals. We also need a way of keeping track of how many role-reversals have occurred.

Definition 3.10 Let φ be a first-order formula, $\mathbb{M}$ a suitable structure and s an assignment in $\mathbb{M}$ whose domain contains Free(φ). The *semantic game* $G(\mathbb{M}, s, \varphi)$ is defined as before, except for the following changes.

history $(\varnothing, \varphi)$, i.e., we wrote $\sigma(\varnothing, \varphi)$ instead of $\sigma((\varnothing, \varphi))$. We will frequently repeat this abuse of notation.

- If ψ is $\neg\chi$, then $H_\chi = \{\, h^\frown\chi : h \in H_{\neg\chi} \,\}$. We can tell which player is the verifier in a history by counting transitions of the form $\neg\chi$ to χ. If there is an even number of such transitions, then Eloise is the verifier; if there is an odd number, then Abelard is the verifier.
- Disjunctions and existential quantifiers are decision points for the verifier p; conjunctions and universal quantifiers are decision points for the falsifier $\bar{p}$:

$$P(h) = \begin{cases} p & \text{if } h \in H_{\chi\vee\chi'} \text{ or } h \in H_{\exists x\chi}, \\ \bar{p} & \text{if } h \in H_{\chi\wedge\chi'} \text{ or } h \in H_{\forall x\chi}. \end{cases}$$

- The verifier p wins a maximal play $h \in H_\chi$ if the atomic formula χ is satisfied by the current assignment. The falsifier $\bar{p}$ wins if it is not:

$$u(h) = \begin{cases} p & \text{if } \mathbb{M}, s_h \models \chi, \\ \bar{p} & \text{if } \mathbb{M}, s_h \not\models \chi. \end{cases} \dashv$$

For example, consider the semantic game $G(\mathbb{N}, \varnothing, \neg\varphi)$, where φ is the sentence $\exists x\forall y(y \leq x)$. Eloise has a winning strategy, namely

$$\sigma\big(\varnothing, \neg\varphi, \varphi, (x, a)\big) = (y, a+1),$$

which is identical to Abelard's strategy for $G(\mathbb{N}, \varnothing, \varphi)$ except for the presence of $\neg\varphi$ in every history. It should be clear from this example that Eloise has a winning strategy for $G(\mathbb{M}, s, \neg\varphi)$ if and only if Abelard has a winning strategy for $G(\mathbb{M}, s, \varphi)$, and vice versa.

3.3.2 Truth and satisfaction

We are now ready to say when a first-order formula is true. The attentive reader should be able to anticipate the definition.

Definition 3.11 Let φ be a first-order formula, $\mathbb{M}$ a suitable structure and s an assignment in $\mathbb{M}$ whose domain contains $\text{Free}(\varphi)$. Then

$$\mathbb{M}, s \models \varphi \quad \text{iff} \quad \text{Eloise has a winning strategy for } G(\mathbb{M}, s, \varphi),$$

in which case we say $(\mathbb{M}, s)$ *satisfies* φ. When φ is a sentence,

$$\mathbb{M} \models \varphi \quad \text{iff} \quad \mathbb{M}, \varnothing \models \varphi.$$

In such cases we say $\mathbb{M}$ *models* φ, and that φ is *true* in $\mathbb{M}$. $\dashv$

Since Eloise has a winning strategy for $G(\mathbb{M}, s, \neg\varphi)$ if and only if Abelard has a winning strategy for $G(\mathbb{M}, s, \varphi)$, it follows that

$$\mathbb{M}, s \models \neg\varphi \quad \text{iff} \quad \text{Abelard has a winning strategy for } G(\mathbb{M}, s, \varphi).$$

We say a sentence φ is *false* in $\mathbb{M}$ when $\mathbb{M} \models \neg\varphi$.

We use the notation $\mathbb{M}, s \not\models \varphi$ when Eloise does *not* have a winning strategy for $G(\mathbb{M}, s, \varphi)$. In such situations we say that $(\mathbb{M}, s)$ does not satisfy φ. When φ is a sentence we write $\mathbb{M} \not\models \varphi$ if $\mathbb{M}, \varnothing \not\models \varphi$ and say $\mathbb{M}$ does not model φ, or that φ is not true in $\mathbb{M}$. We wish to emphasize that it is conceivable for Eloise not to have a winning strategy for a particular semantic game without Abelard having a winning strategy, either. Fear not, gentle reader! The Gale-Stewart theorem will come to the rescue.

Proposition 3.12 *Let φ be a first-order formula, $\mathbb{M}$ a suitable structure, and s an assignment in $\mathbb{M}$ whose domain contains* Free(φ). *Then*

$$\mathbb{M}, s \models \neg\varphi \quad \text{iff} \quad \mathbb{M}, s \not\models \varphi.$$

Proof $G(\mathbb{M}, s, \varphi)$ is a two-player, win-lose game with finite horizon and a unique initial history. If Abelard has a winning strategy, then Eloise does not (because the game is win-lose). Conversely, if Eloise does not have a winning strategy, then Abelard must have one by the Gale-Stewart theorem. ⊣

It follows that, when φ is a sentence, $\mathbb{M} \models \neg\varphi$ if and only if $\mathbb{M} \not\models \varphi$.

The values of free variables are all that matter when determining whether an assignment satisfies a formula in a given structure. If s assigns values to x, y and z, we can ignore the value of z when evaluating formulas such as $R(x, y)$ or $\exists z(\varphi \wedge \psi)$.

Proposition 3.13 *Let φ be a first-order formula, $\mathbb{M}$ a suitable structure, and s, s' assignments in $\mathbb{M}$ that agree on* Free(φ). *Then*

$$\mathbb{M}, s \models \varphi \quad \text{iff} \quad \mathbb{M}, s' \models \varphi.$$

Proof Suppose Eloise has a winning strategy σ for $G = G(\mathbb{M}, s, \varphi)$. Since every history $h = (s, \varphi, \ldots)$ for G corresponds to a history $h' = (s', \varphi, \ldots)$ for $G' = G(\mathbb{M}, s', \varphi)$ obtained by substituting s' for s and leaving the rest of the history unchanged, we can define a strategy σ' for G' by $\sigma'(h') = \sigma(h)$. That is, σ' tells Eloise to mimic her winning strategy for G in the game G'.

Now suppose $h' = (s', \varphi, \ldots, \chi)$ is a terminal history for G' in which Eloise follows σ'. Then $h = (s, \varphi, \ldots, \chi)$ is a terminal history for the game $G(\mathbb{M}, s, \varphi)$ in which she follows σ. A simple induction shows that the assignments s_h and $s_{h'}$ agree on Free(χ). Therefore Eloise wins h' if and only if she wins h, which she does because σ is a winning strategy. Thus σ' is a winning strategy for G'. The converse is symmetrical. ⊣

A consequence of Proposition 3.13 is that we can play semantic games without remembering every move. For example, if a variable is quantified twice, as in the formula $\forall x \forall x \exists y (x = y)$, Abelard chooses the value of x twice, but only his second choice matters. Eloise need only consider the "current" value of x when picking the value of y.

In the semantic game for a first-order formula, the state of play is encoded by an assignment and the current subformula.

Definition 3.14 A strategy σ for $G(\mathbb{M}, s, \varphi)$ is *memoryless* if for every history h, the action $\sigma(h)$ only depends on the current assignment and the current subformula, i.e., for every nonatomic subformula $\psi \in \text{Subf}(\varphi)$, if $h, h' \in H_\psi$ and $s_h = s_{h'}$, then $\sigma(h) = \sigma(h')$. ⊣

Proposition 3.15 *If a player has a winning strategy for the semantic game of a first-order formula φ, then he or she has a memoryless winning strategy.*

Proof Suppose player p has a winning strategy σ for $G(\mathbb{M}, s, \varphi)$. If φ is atomic, then σ is the empty strategy, which is memoryless. If φ is $\neg\psi$ the opponent $\bar{p}$ has a winning strategy τ for $G(\mathbb{M}, s, \psi)$ given by

$$\tau(s, \psi, \ldots) = \sigma(s, \neg\psi, \psi, \ldots).$$

That is, $\tau(h) = \sigma(h')$, where h' is the history for $G(\mathbb{M}, s, \neg\psi)$ that is identical to h except for the insertion of $\neg\psi$ after the initial assignment. By inductive hypothesis, $\bar{p}$ has a memoryless winning strategy τ' for $G(\mathbb{M}, s, \psi)$. Hence p has a memoryless winning strategy for $G(\mathbb{M}, s, \neg\psi)$ given by

$$\sigma'(s, \neg\psi, \psi, \ldots) = \tau'(s, \psi, \ldots).$$

From now on we will assume p is Eloise.

The inductive step for disjunctions and conjunctions is straightforward. If φ is $\exists x \psi$, let $\sigma(x, \exists x \psi) = (x, a)$, and define

$$\sigma'\big(s(x/a), \psi, \ldots\big) = \sigma\big(s, \exists x \psi, (x, a), \ldots\big).$$

Then σ' is winning for $G\big(\mathbb{M}, s(x/a), \psi\big)$, so by inductive hypothesis Eloise has a memoryless winning strategy σ'' for $G\big(\mathbb{M}, s(x/a), \psi\big)$. Hence the strategy σ''' defined by

$$\sigma'''(s, \exists x\psi) = (x, a),$$
$$\sigma'''(s, \exists x\psi, (x, a), \ldots) = \sigma''(s(x/a), \psi, \ldots),$$

is a memoryless winning strategy for Eloise in $G(\mathbb{M}, s, \exists x\psi)$.

Finally, if φ is $\forall x\psi$ then, for all $a \in M$, Eloise has a winning strategy σ_a for $G(\mathbb{M}, s(x/a), \psi)$ defined by

$$\sigma_a(s(x/a), \psi, \ldots) = \sigma(s, \forall x\psi, (x, a), \ldots).$$

By inductive hypothesis she has memoryless winning strategies σ'_a, as well. Define a strategy σ' by

$$\sigma'(s, \forall x\psi, (x, a), \ldots) = \sigma'_a(s(x/a), \psi, \ldots).$$

Then σ' is a memoryless winning strategy for Eloise in $G(\mathbb{M}, s, \forall x\psi)$. ⊣

3.4 Logical equivalence

A player may have multiple winning strategies for a semantic game. It may also happen that essentially the same strategy is winning for two different games. Two games may be so similar, in fact, that every winning strategy for one induces a winning strategy for the other, and vice versa.

Definition 3.16 Let φ and ψ be first-order formulas. We say that φ *entails* ψ, denoted $\varphi \models \psi$, if for every suitable structure $\mathbb{M}$ and assignment s,

$$\mathbb{M}, s \models \varphi \quad \text{implies} \quad \mathbb{M}, s \models \psi.$$

We say that φ and ψ are *equivalent*, denoted $\varphi \equiv \psi$, if $\varphi \models \psi$ and $\psi \models \varphi$. ⊣

One can easily prove the standard laws of propositional logic using game-theoretic semantics. Let $\top$ be an abbreviation for the sentence $\forall x(x = x)$, and let $\bot$ be an abbreviation for $\exists x(x \neq x)$.

Proposition 3.17 *Let φ, ψ, and χ be first-order formulas.*

Commutativity
$$\varphi \vee \psi \equiv \psi \vee \varphi.$$
$$\varphi \wedge \psi \equiv \psi \wedge \varphi.$$

Associativity

$$\varphi \vee (\psi \vee \chi) \equiv (\varphi \vee \psi) \vee \chi.$$
$$\varphi \wedge (\psi \wedge \chi) \equiv (\varphi \wedge \psi) \wedge \chi.$$

Absorption

$$\varphi \vee (\varphi \wedge \psi) \equiv \varphi.$$
$$\varphi \wedge (\varphi \vee \psi) \equiv \varphi.$$

Distributivity

$$\varphi \vee (\psi \wedge \chi) \equiv (\varphi \vee \psi) \wedge (\varphi \vee \chi).$$
$$\varphi \wedge (\psi \vee \chi) \equiv (\varphi \wedge \psi) \vee (\varphi \wedge \chi).$$

Complementation

$$\varphi \vee \neg\varphi \equiv \top.$$
$$\varphi \wedge \neg\varphi \equiv \bot.$$

Proof We only prove the first of each pair of logical equivalences.

(Commutativity) Suppose Eloise has a winning strategy σ for $G = G(\mathbb{M}, s, \varphi \vee \psi)$. Define a strategy σ' for $G' = G(\mathbb{M}, s, \psi \vee \varphi)$ by

$$\sigma'(s, \psi \vee \varphi, \ldots) = \sigma(s, \varphi \vee \psi, \ldots).$$

That is, if σ tells Eloise to choose the left disjunct, then σ' tells her to choose the right disjunct, and vice versa. Every terminal history $h' = (s, \psi \vee \varphi, \chi, \ldots)$ for G' in which Eloise follows σ' corresponds to a history $h = (s, \varphi \vee \psi, \chi, \ldots)$ for G in which she follows σ that induces the same assignment and terminates at the same atomic subformula as h'. Thus Eloise wins h' if and only if she wins h, which she does because σ is winning strategy for G. Therefore σ' is a winning strategy for G'. The converse is symmetrical.

(Associativity) Suppose Eloise has a winning strategy σ for $G = G(\mathbb{M}, s, \varphi \vee (\psi \vee \chi))$. Define a strategy σ' for $G' = G(\mathbb{M}, s, (\varphi \vee \psi) \vee \chi)$ as follows. If $\sigma(s, \varphi \vee (\psi \vee \chi)) = \varphi$, then

$$\sigma'(s, (\varphi \vee \psi) \vee \chi) = (\varphi \vee \psi),$$
$$\sigma'(s, (\varphi \vee \psi) \vee \chi, (\varphi \vee \psi)) = \varphi,$$
$$\sigma'(s, (\varphi \vee \psi) \vee \chi, (\varphi \vee \psi), \varphi, \ldots) = \sigma(s, \varphi \vee (\psi \vee \chi), \varphi, \ldots).$$

If $\sigma(s, \varphi \vee (\psi \vee \chi)) = (\psi \vee \chi)$ and $\sigma(s, \varphi \vee (\psi \vee \chi), (\psi \vee \chi)) = \psi$, then

$$\sigma'(s, (\varphi \vee \psi) \vee \chi) = (\varphi \vee \psi),$$
$$\sigma'(s, (\varphi \vee \psi) \vee \chi, (\varphi \vee \psi)) = \psi,$$
$$\sigma'(s, (\varphi \vee \psi) \vee \chi, (\varphi \vee \psi), \psi, \ldots) = \sigma(s, \varphi \vee (\psi \vee \chi), (\psi \vee \chi), \psi, \ldots).$$

If $\sigma\big(s, \varphi \vee (\psi \vee \chi)\big) = (\psi \vee \chi)$ and $\sigma\big(s, \varphi \vee (\psi \vee \chi), (\psi \vee \chi)\big) = \chi$, then

$$\sigma'\big(s, (\varphi \vee \psi) \vee \chi\big) = \chi,$$
$$\sigma'\big(s, (\varphi \vee \psi) \vee \chi, \chi, \ldots\big) = \sigma\big(s, \varphi \vee (\psi \vee \chi), (\psi \vee \chi), \chi, \ldots\big).$$

That is, σ' tells Eloise to choose the same disjunct (φ, ψ, or χ) as σ, and to mimic σ thereafter.

In the first case, every terminal history

$$h' = \big(s, (\varphi \vee \psi) \vee \chi, (\varphi \vee \psi), \varphi, \ldots\big)$$

for G' in which Eloise follows σ' corresponds to a terminal history

$$h = \big(s, \varphi \vee (\psi \vee \chi), \varphi, \ldots\big)$$

for G in which she follows σ that induces the same assignment and terminates at the same atomic formula as h'. Thus Eloise wins h' if and only if she wins h, which she does because σ is a winning strategy for G. Therefore σ' is a winning strategy for G'.

The other two cases are similar, and the converse is symmetrical.

(Absorption) Suppose that Eloise has a winning strategy σ for $G = G(\mathbb{M}, s, \varphi)$. Define a strategy σ' for $G' = G\big(\mathbb{M}, s, \varphi \vee (\varphi \wedge \psi)\big)$ by:

$$\sigma'\big(s, \varphi \vee (\varphi \wedge \psi)\big) = \varphi,$$
$$\sigma'\big(s, \varphi \vee (\varphi \wedge \psi), \varphi, \ldots\big) = \sigma(s, \varphi, \ldots).$$

That is, σ' tells Eloise to choose φ, then mimic σ thereafter. Every terminal history $h' = \big(s, \varphi \vee (\varphi \wedge \psi), \varphi, \ldots\big)$ for G' in which Eloise follows σ' corresponds to a terminal history $h = (s, \varphi, \ldots)$ for G in which Eloise follows σ. Therefore Eloise wins h' if and only if she wins h, which she does because σ is a winning strategy for G. Thus σ' is a winning strategy for G'.

Conversely, suppose Eloise has a winning strategy σ' for G'. Define a winning strategy σ for G as follows. If $\sigma'\big(s, \varphi \vee (\varphi \wedge \psi)\big) = \varphi$, let

$$\sigma(s, \varphi, \ldots) = \sigma'\big(s, \varphi \vee (\varphi \wedge \psi), \varphi, \ldots\big).$$

If $\sigma'\big(s, \varphi \vee (\varphi \wedge \psi)\big) = (\varphi \wedge \psi)$, let

$$\sigma(s, \varphi, \ldots) = \sigma'\big(s, \varphi \vee (\varphi \wedge \psi), (\varphi \wedge \psi), \varphi, \ldots\big).$$

That is, if σ' tells Eloise to choose φ, then σ tells her to mimic σ'. If σ' tells Eloise to choose $(\varphi \wedge \psi)$, then σ tells her to mimic the part of σ' she follows when Abelard chooses φ.

(Distributivity) The proof of the distributive laws is similar. The reader should test his or her mastery of game-theoretic semantics by writing out all the details.

(Complementation) If Eloise has a winning strategy for $G(\mathbb{M}, s, \varphi)$ she can extend it to a winning strategy for $G(\mathbb{M}, s, \varphi \vee \neg\varphi)$ by always choosing the left disjunct. If she does not have a winning strategy for $G(\mathbb{M}, s, \varphi)$, then Abelard does, which implies Eloise has a winning strategy for $G(\mathbb{M}, s, \neg\varphi)$. Eloise can extend her strategy to a winning strategy for $G(\mathbb{M}, s, \varphi \vee \neg\varphi)$ by always choosing the right disjunct.   ⊣

The game-theoretic perspective allows us to see the true import of classical laws such as double negation and De Morgan's laws — they are statements about which player is the verifier and whose turn it is to move. Since a negated formula tells the players to switch roles, a doubly negated formula tells the players to switch roles twice. It would be analogous to a rule in chess that tells the players to switch colors. Applied twice in succession the rule would tell White to continue playing as White, and Black as Black.

When Eloise denies a conjunction, she may pick which conjunct to falsify. On the other hand, if Eloise denies a disjunction, Abelard decides which disjunct she must falsify. Similarly, when Eloise denies a universal formula $\forall x \varphi$, she may choose the value of x before attempting to falsify φ. If she denies an existential formula $\exists x \varphi$, Abelard chooses the value of x.

Proposition 3.18   *Let φ and ψ be first-order formulas.*

(a) $\neg\neg\varphi \equiv \varphi$.

(b) $\neg(\varphi \wedge \psi) \equiv \neg\varphi \vee \neg\psi$.
     $\neg(\varphi \vee \psi) \equiv \neg\varphi \wedge \neg\psi$.

(c) $\neg\forall x \varphi \equiv \exists x(\neg\varphi)$.
     $\neg\exists x \varphi \equiv \forall x(\neg\varphi)$.

Proof   (a) Eloise has a winning strategy for $G(\mathbb{M}, s, \neg\neg\varphi)$ if and only if Abelard has a winning strategy for $G(\mathbb{M}, s, \neg\varphi)$ if and only if Eloise has a winning strategy for $G(\mathbb{M}, s, \varphi)$.

(b) Suppose Eloise has a winning strategy σ for $G\big(\mathbb{M}, s, \neg(\varphi \wedge \psi)\big)$. Define a winning strategy σ' for Eloise in $G(\mathbb{M}, s, \neg\varphi \vee \neg\psi)$ as follows:

$$\sigma'(s, \neg\varphi \vee \neg\psi) = \begin{cases} \neg\varphi & \text{if } \sigma\big(s, \neg(\varphi \wedge \psi), (\varphi \wedge \psi)\big) = \varphi, \\ \neg\psi & \text{if } \sigma\big(s, \neg(\varphi \wedge \psi), (\varphi \wedge \psi)\big) = \psi, \end{cases}$$

and

$$\sigma'(s, \neg\varphi \vee \neg\psi, \neg\varphi, \varphi, \ldots) = \sigma(s, \neg(\varphi \wedge \psi), (\varphi \wedge \psi), \varphi, \ldots),$$
$$\sigma'(s, \neg\varphi \vee \neg\psi, \neg\psi, \psi, \ldots) = \sigma(s, \neg(\varphi \wedge \psi), (\varphi \wedge \psi), \psi, \ldots).$$

Conversely, suppose Eloise has a winning strategy σ for the semantic game $G(\mathbb{M}, s, \neg\varphi \vee \neg\psi)$. Then she has a winning strategy σ' for the game $G(\mathbb{M}, s, \neg(\varphi \vee \psi))$ defined by

$$\sigma'(s, \neg(\varphi \wedge \psi), (\varphi \wedge \psi)) = \begin{cases} \varphi & \text{if } \sigma(s, \neg\varphi \vee \neg\psi) = \neg\varphi, \\ \psi & \text{if } \sigma(s, \neg\varphi \vee \neg\psi) = \neg\psi, \end{cases}$$

and

$$\sigma'(s, \neg(\varphi \wedge \psi), (\varphi \wedge \psi), \varphi, \ldots) = \sigma(s, \neg\varphi \vee \neg\psi, \neg\varphi, \varphi, \ldots),$$
$$\sigma'(s, \neg(\varphi \wedge \psi), (\varphi \wedge \psi), \psi, \ldots) = \sigma(s, \neg\varphi \vee \neg\psi, \neg\psi, \psi, \ldots).$$

The proof of the dual is similar.

(c) For every strategy σ for Eloise in $G(\mathbb{M}, s, \neg\forall x\varphi)$, let σ' be her strategy for $G(\mathbb{M}, s, \exists x(\neg\varphi))$ defined by

$$\sigma'(s, \exists x(\neg\varphi), \ldots) = \sigma(s, \neg\forall x\varphi, \forall x\varphi, \ldots),$$
$$\sigma'(s, \exists x(\neg\varphi), (x, a), \varphi, \ldots) = \sigma(s, \neg\forall x\varphi, \forall x\varphi, (x, a), \ldots),$$

and vice versa. Then σ is a winning strategy if and only if σ' is too. $\dashv$

Proposition 3.18 allows us to adjust the position of negation symbols in a first-order formula while preserving its meaning. For instance, we can push every negation symbol as deep into the formula as possible, stopping when we reach an atomic formula. A first-order formula φ is in *negation normal form* if the only negated subformulas of φ are atomic. By repeated application of Proposition 3.18 we can show that every first-order formula is equivalent to a formula in negation normal form, which means that we can delay role-reversals until the end of the semantic game.

Game-theoretic semantics also sheds light on the interaction between quantifiers and connectives.

Proposition 3.19 *Let φ and ψ be first-order formulas.*

(a) $\exists x(\varphi \vee \psi) \equiv \exists x\varphi \vee \exists x\psi.$
$\quad\;\; \forall x(\varphi \wedge \psi) \equiv \forall x\varphi \wedge \forall x\psi.$

(b) $\exists x(\varphi \wedge \psi) \models \exists x\varphi \wedge \exists x\psi$.
$\quad \forall x\varphi \vee \forall x\psi \models \forall x(\varphi \vee \psi)$.

Proof (a) Suppose Eloise has a winning strategy σ for the game $G = G(\mathbb{M}, s, \exists x(\varphi \vee \psi))$. Let

$$\sigma(s, \exists x(\varphi \vee \psi)) = (x, a),$$
$$\sigma(s, \exists x(\varphi \vee \psi), (x, a)) = \chi.$$

Define a strategy σ' for $G' = G(\mathbb{M}, s, \exists x\varphi \vee \exists x\psi)$ as follows:

$$\sigma'(s, \exists x\varphi \vee \exists x\psi) = \exists x\chi,$$
$$\sigma'(s, \exists x\varphi \vee \exists x\psi, \exists x\chi) = (x, a),$$
$$\sigma'(s, \exists x\varphi \vee \exists x\psi, \exists x\chi, (x, a), \ldots) = \sigma(s, \exists x(\varphi \vee \psi), (x, a), \chi, \ldots).$$

That is, σ' tells Eloise to choose $\exists x\varphi$ if she picks φ in G, to choose $\exists x\psi$ if she picks ψ, and to assign x the same value as she did in G.

Observe that in both games, after Eloise's first two moves, the current assignment is $s(x/a)$ and the current subformula is χ. Thus play proceeds as in the game $G(\mathbb{M}, s(x/a), \chi)$. Every terminal history

$$h' = (s, \exists x\varphi \vee \exists x\psi, \exists x\chi, (x, a), \ldots)$$

of G' in which Eloise follows σ' corresponds to a terminal history

$$h = (s, \exists x(\varphi \vee \psi), (x, a), \chi, \ldots).$$

of G in which Eloise follows σ that induces the same assignment and terminates at the same atomic formula as h'. Thus Eloise wins h' if and only if she wins h, which she does because σ is a winning strategy. Therefore σ' is a winning strategy for G'.

The proof of the converse is symmetrical.

(b) Let $G = G(\mathbb{M}, s, \exists x(\varphi \wedge \psi))$ and $G' = G(\mathbb{M}, s, \exists x\varphi \wedge \exists x\psi)$. Suppose Eloise has a winning strategy σ for G. Let $\sigma(s, \exists x(\varphi \wedge \psi)) = (x, a)$. Define a strategy σ' for G' by:

$$\sigma'(s, \exists x\varphi \wedge \exists x\psi, \exists x\varphi) = (x, a),$$
$$\sigma'(s, \exists x\varphi \wedge \exists x\psi, \exists x\varphi, (x, a), \ldots) = \sigma(s, \exists x(\varphi \wedge \psi), (x, a), \varphi, \ldots),$$

and

$$\sigma'(s, \exists x\varphi \wedge \exists x\psi, \exists x\psi) = (x, a),$$
$$\sigma'(s, \exists x\varphi \wedge \exists x\psi, \exists x\psi, (x, a), \ldots) = \sigma(s, \exists x(\varphi \wedge \psi), (x, a), \psi, \ldots).$$

That is, σ' tells Eloise to choose a for the value of x no matter which conjunct Abelard chooses, and to mimic σ thereafter.

Observe that in both games, after each player makes their first move the current assignment is $s(x/a)$, and the current subformula is φ or ψ. In either case, every terminal history h' for G' in which Eloise follows σ' corresponds to a terminal history h for G in which Eloise follows σ that induces the same assignment and terminates at the same atomic subformula as h'. Thus Eloise wins h' if and only if she wins h, which she does because σ is a winning strategy for G. Therefore σ' is a winning strategy for G'. ⊣

We can now see that existential quantifiers distribute over disjunctions because they are both moves for the same player, whereas existential quantifiers fail to distribute over conjunctions because they are moves for different players. In the first case, Eloise can plan ahead and choose the value of x that will verify the appropriate disjunct, or choose the disjunct first and then choose the value of x. In the second case, she is forced to commit to a value of x before she knows which conjunct Abelard will choose. Universal quantifiers distribute over conjunctions, but not disjunctions, for the same reason.

A similar argument shows that adjacent existential quantifiers commute, as do adjacent universal quantifiers. If a player is asked to choose the value of several variables in a row, the order in which he or she chooses them does not matter. Existential quantifiers do not commute with universal quantifiers, however. We leave the proof of the next proposition as an exercise for the reader.

Proposition 3.20 *Let φ and ψ be first-order formulas.*

(a) $\exists x \exists y \varphi \equiv \exists y \exists x \varphi.$
   $\forall x \forall y \varphi \equiv \forall y \forall x \varphi.$
(b) $\exists x \forall y \varphi \models \forall y \exists x \varphi.$ ⊣

The failure of the converses of Proposition 3.19(b) and Proposition 3.20(b) shows that the order in which the players make their moves can affect whether a player has a winning strategy. The value of a variable only affects subformulas in which that variable is free, so the scope of a quantifier Qx may be expanded to include subformulas in which x is not free.

Proposition 3.21 *Let φ and ψ be first-order formulas such that $x \notin$ Free(ψ).*

(a) $\exists x(\varphi \vee \psi) \equiv \exists x\varphi \vee \psi.$
  $\forall x(\varphi \wedge \psi) \equiv \forall x\varphi \wedge \psi.$
(b) $\exists x(\varphi \wedge \psi) \equiv \exists x\varphi \wedge \psi.$
  $\forall x(\varphi \vee \psi) \equiv \forall x\varphi \vee \psi.$

Proof (a) Let $G = G(\mathbb{M}, s, \exists x(\varphi \vee \psi))$ and $G' = G(\mathbb{M}, s, \exists x\varphi \vee \psi)$. Suppose Eloise has a winning strategy σ for G such that

$$\sigma\big(s, \exists x(\varphi \vee \psi)\big) = (x, a).$$

Define a strategy σ' for G' as follows. If $\sigma\big(s, \exists x(\varphi \vee \psi), (x,a)\big) = \varphi$, let

$$\sigma'(s, \exists x\varphi \vee \psi) = \exists x\varphi,$$
$$\sigma'(s, \exists x\varphi \vee \psi, \exists x\varphi) = (x, a),$$
$$\sigma'\big(s, \exists x\varphi \vee \psi, \exists x\varphi, (x,a), \ldots\big) = \sigma\big(s, \exists x(\varphi \vee \psi), (x,a), \varphi, \ldots\big).$$

That is, Eloise chooses the left disjunct, sets the value of x to a, then mimics σ thereafter. If $\sigma\big(s, \exists x(\varphi \vee \psi), (x,a)\big) = \psi$, let

$$\sigma'(s, \exists x\varphi \vee \psi) = \psi,$$
$$\sigma'\big(s, \exists x\varphi \vee \psi, \psi, \ldots\big) = \sigma\big(s, \exists x(\varphi \vee \psi), (x,a), \psi, \ldots\big).$$

That is, Eloise chooses the right disjunct, then mimics σ thereafter.
  In the first case, every terminal history

$$h' = \big(s, \exists x\varphi \vee \psi, \exists x\varphi, (x,a), \ldots\big)$$

in which Eloise follows σ' corresponds to a history

$$h = \big(s, \exists x(\varphi \vee \psi), (x,a), \varphi, \ldots\big)$$

in which Eloise follows σ that induces the same assignment and terminates at the same atomic subformula as h'. Thus Eloise wins h' if and only if she wins h, which she does because σ is a winning strategy for G. In the second case, every terminal history

$$h' = (s, \exists x\varphi \vee \psi, \psi, \ldots,)$$

in which Eloise follows σ' corresponds to a history

$$h = \big(s, \exists x(\varphi \vee \psi), (x,a), \psi, \ldots\big)$$

in which Eloise follows σ that terminates at the same atomic subformula χ and induces an assignment that differs from $s_{h'}$ on at most the value of x. If x does not occur in χ, then its value does not matter. If x does occur in χ, it can only be because ψ has a subformula $Qx\psi'$ of which

χ is a proper subformula. Hence the value of x was reset at some point during play, i.e.,

$$h = \big(s, \exists x(\varphi \vee \psi), (x,a), \psi, \dots, (x,b), \dots\big),$$
$$h' = \big(s, \exists x \varphi \vee \psi, \psi, \dots, (x,b), \dots\big).$$

Thus $s_h = s_{h'}$, so Eloise wins both h and h'. Therefore σ' is a winning strategy for G'.

The proof of the converse is similar.

(b) Let $G = G(\mathbb{M}, s, \exists x \varphi \wedge \psi)$ and $G' = G\big(\mathbb{M}, s, \exists x(\varphi \wedge \psi)\big)$. Suppose Eloise has a winning strategy σ for G such that $\sigma(s, \exists x \varphi \wedge \psi, \exists x \varphi) = (x,a)$. Define a strategy σ' for G' as follows:

$$\sigma'\big(s, \exists x(\varphi \wedge \psi)\big) = (x,a),$$
$$\sigma'\big(s, \exists x(\varphi \wedge \psi), (x,a), \varphi, \dots\big) = \sigma\big(s, \exists x \varphi \wedge \psi, \exists x \varphi, (x,a), \dots\big),$$
$$\sigma'\big(s, \exists x(\varphi \wedge \psi), (x,a), \psi, \dots\big) = \sigma\big(\exists x \varphi \wedge \psi, \psi, \dots\big).$$

On the one hand, for every terminal history

$$h' = (s, \exists x(\varphi \wedge \psi), (x,a), \varphi, \dots)$$

for G' in which Eloise follows σ' corresponds to a terminal history

$$h = (s, \exists x \varphi \wedge \psi, \exists x \varphi, (x,a), \dots)$$

for G in which Eloise follows σ that induces the same assignment and terminates at the same atomic subformula as h'. Thus Eloise wins h' if and only if she wins h, which she does because σ is a winning strategy. On the other hand, for every terminal history

$$h' = (s, \exists x(\varphi \wedge \psi), (x,a), \psi, \dots)$$

for G' in which Eloise follows σ' there is a corresponding terminal history

$$h = (s, \exists x \varphi \wedge \psi, \psi, \dots)$$

for G in which Eloise follows σ that terminates at the same atomic subformula χ. The two histories may not induce the same assignment, but they do agree on the value of every variable other than x. If x does not occur in χ, then its value does not matter. Eloise wins h' if and only if she wins h, which she does because σ is a winning strategy. If x does occur in χ, then it can only be because ψ has a subformula $Qx\psi'$ of which χ is a proper subformula. Hence the value of x was reset at some point during play. That is,

$$h = (s, \exists x\varphi \wedge \psi, \psi, \ldots, (x, b), \ldots),$$
$$h' = (s, \exists x(\varphi \wedge \psi), (x, a), \psi, \ldots, (x, b), \ldots).$$

Thus $s_h = s_{h'}$ so Eloise wins both h and h'. Therefore σ' is a winning strategy for G'.   ⊣

3.5 Compositional semantics

Now that the game-theoretic nature of first-order logic has been firmly established, one can ask if there is a method that determines the semantic value of a formula in a particular model out of the semantic values of its subformulas. Of course, the answer is yes. Tarski discovered such a method and proposed using it as the *definition* of truth before game-theoretic semantics was fully developed. We believe the game-theoretic account is an improvement over Tarski's recursive definition because it explains certain logical phenomena (such as the duality of $\exists$ and $\forall$) that the recursive definition merely asserts. That is, in the Tarskian tradition it is common to simply define $\forall x\varphi$ as $\neg\exists x(\neg\varphi)$. Of course one can define both $\exists x$ and $\forall x$ and prove their duality, but the proof is not very explanatory because it depends on the duality of "there exists" and "for all" in English.

Tarski's method remains exceedingly useful, nonetheless. For example, we will use it to prove that every first-order formula is equivalent to one in prenex normal form (see Theorem 3.26).

Theorem 3.22 *Let φ and ψ be first-order formulas, $\mathbb{M}$ a suitable structure, and s an assignment in $\mathbb{M}$ whose domain contains* Free(φ) *and* Free(ψ).

$$\mathbb{M}, s \models \neg\varphi \quad \textit{iff} \quad \mathbb{M}, s \not\models \varphi.$$
$$\mathbb{M}, s \models \varphi \vee \psi \quad \textit{iff} \quad \mathbb{M}, s \models \varphi \textit{ or } \mathbb{M}, s \models \psi.$$
$$\mathbb{M}, s \models \varphi \wedge \psi \quad \textit{iff} \quad \mathbb{M}, s \models \varphi \textit{ and } \mathbb{M}, s \models \psi.$$
$$\mathbb{M}, s \models \exists x\varphi \quad \textit{iff} \quad \mathbb{M}, s(x/a) \models \varphi, \textit{ for some } a \in M.$$
$$\mathbb{M}, s \models \forall x\varphi \quad \textit{iff} \quad \mathbb{M}, s(x/a) \models \varphi, \textit{ for every } a \in M.$$

Proof We have already proven the negation case in Proposition 3.12. Suppose Eloise has a winning strategy σ for $G(\mathbb{M}, s, \varphi \vee \psi)$ such that

$\sigma(s, \varphi \vee \psi) = \chi$. Define a strategy σ' for $G(\mathbb{M}, s, \chi)$ by

$$\sigma'(s, \chi, \ldots) = \sigma(s, \varphi \vee \psi, \chi, \ldots).$$

That is, σ' is the strategy for $G(\mathbb{M}, s, \chi)$ that mimics the part of σ Eloise consults after choosing χ. Every terminal history $h' = (s, \chi, \ldots)$ for $G(\mathbb{M}, s, \chi)$ in which Eloise follows σ' corresponds to a history $h = (s, \varphi \vee \psi, \chi, \ldots)$ in which she follows σ that induces the same assignment and terminates at the same atomic subformula as h'. Thus Eloise wins h' if and only if she wins h, which she does because σ is a winning strategy for $G(\mathbb{M}, s, \varphi \vee \psi)$. Therefore σ' is a winning strategy for $G(\mathbb{M}, s, \chi)$.

Conversely, let $\chi \in \{\varphi, \psi\}$, and suppose Eloise has a winning strategy σ' for $G(\mathbb{M}, s, \chi)$. Define a winning strategy σ for $G(\mathbb{M}, s, \varphi \vee \psi)$ by

$$\sigma(s, \varphi \vee \psi) = \chi,$$
$$\sigma(s, \varphi \vee \psi, \chi, \ldots) = \sigma'(s, \chi, \ldots).$$

That is, σ tells Eloise to choose the disjunct for which she has a winning strategy, then follow that strategy.

Now suppose Eloise has a winning strategy σ for $G(\mathbb{M}, s, \varphi \wedge \psi)$. Define

$$\sigma'(s, \varphi, \ldots) = \sigma(s, \varphi \wedge \psi, \varphi, \ldots),$$
$$\sigma''(s, \psi, \ldots) = \sigma(s, \varphi \wedge \psi, \psi, \ldots).$$

That is, σ' is the part of σ that Eloise consults when Abelard chooses φ, and σ'' is the part of σ she consults when he chooses ψ. Then σ' is a winning strategy for $G(\mathbb{M}, s, \varphi)$, and σ'' is a winning strategy for $G(\mathbb{M}, s, \psi)$. Conversely, suppose Eloise has winning strategies σ' for $G(\mathbb{M}, s, \varphi)$ and σ'' for $G(\mathbb{M}, s, \psi)$. Define

$$\sigma(s, \varphi \wedge \psi, \varphi, \ldots) = \sigma'(s, \varphi, \ldots),$$
$$\sigma(s, \varphi \wedge \psi, \psi, \ldots) = \sigma''(s, \psi, \ldots).$$

That is, σ is the winning strategy for $G(\mathbb{M}, s, \varphi \wedge \psi)$ that tells Eloise to mimic σ' if Abelard chooses φ and to mimic σ'' if he chooses ψ.

Suppose Eloise has a winning strategy σ for $G(\mathbb{M}, s, \exists x \varphi)$ such that $\sigma(s, \exists x \varphi) = (x, a)$. Then the strategy σ' for $G(\mathbb{M}, s(x/a), \varphi)$ defined by

$$\sigma'(s(x/a), \varphi, \ldots) = \sigma(s, \exists x \varphi, (x, a), \ldots)$$

is winning. Conversely, suppose there exists an $a \in M$ such that Eloise has a winning strategy σ' for $G(\mathbb{M}, s(x/a), \varphi)$. Then the strategy σ for

$G(\mathbb{M}, s, \exists x \varphi)$ defined by

$$\sigma(s, \exists x \varphi) = (x, a),$$
$$\sigma\big(s, \exists x \varphi, (x, a), \ldots\big) = \sigma'\big(s(x/a), \varphi, \ldots\big),$$

is winning.

Finally, suppose Eloise has a winning strategy σ for $G(\mathbb{M}, s, \forall x \varphi)$. For every $a \in M$, define

$$\sigma_a\big(s(x/a), \varphi, \ldots\big) = \sigma\big(s, \forall x \varphi, (x, a), \ldots\big).$$

That is, σ_a tells Eloise to mimic the part of σ she consults when Abelard chooses a. Observe that σ_a is winning for $G\big(\mathbb{M}, s(x/a), \varphi\big)$. Conversely, suppose that for every $a \in M$, Eloise has a winning strategy σ_a for $G\big(\mathbb{M}, s(x/a), \varphi\big)$. Define a winning strategy for $G(\mathbb{M}, s, \forall x \varphi)$ by

$$\sigma\big(s, \forall x \varphi, (x, a), \ldots\big) = \sigma_a\big(s(x/a), \varphi, \ldots\big).$$

That is, σ tells Eloise to mimic σ_a when Abelard sets the value of x to be a.   ⊣

3.5.1 Substitution

The operation of substituting terms for free variables will be used extensively throughout this book. If φ is a quantifier-free formula, x is a variable, and t is a term, $\mathrm{Subst}(\varphi, x, t)$ denotes the first-order formula obtained from φ by simultaneously replacing all free occurrences of x in φ by the term t. For example,

$$\mathrm{Subst}\big(R(x, y), x, u^2\big) \quad \text{is} \quad R(u^2, y),$$

while $\mathrm{Subst}\big(R(x, y), z, u^2\big)$ is simply $R(x, y)$.

Substitution involving formulas with both free and bound variables is a more delicate matter than one might expect. Consider the formula

$$\exists z (z + z = x).$$

Interpreted in the natural numbers, it asserts that x is even. Similarly, if we substitute y for x, the resulting formula

$$\exists z (z + z = y)$$

asserts that y is even. However, if we substitute z for x, we obtain

$$\exists z(z + z = z),$$

which does not does not assert anything at all about z — it is simply true.[4]

The meaning of the formula changed because the variable z became bound by the quantifier $\exists z$. To avoid this problem, whenever we substitute a term for a free variable that lies within the scope of a quantifier that binds a variable occurring in the term, we first rename the bound variable using a fresh variable that occurs nowhere in the formula nor the term. That way, when we substitute z for x in the formula $\exists z(z + z = x)$ we obtain

$$\exists u(u + u = z),$$

which asserts that z is even.

If the substituted term contains more than one variable that would become bound after performing the substitution, we must rename all of those variables bound in the formula simultaneously, taking care to keep distinct variables distinct. For example, when we substitute $f(x, y)$ for z in the formula $\forall x \exists y(x < y \land y < z)$ we obtain

$$\forall u \exists v\big(u < v \land v < f(x, y)\big).$$

Definition 3.23 If x is a variable and t is a term, we can substitute t for x in another term t' by simultaneously replacing every occurrence of x in t' with an occurrence of t. More formally, if c is a constant symbol, y is a variable distinct from x, f is an n-ary function symbol, and $t'_1, \ldots, t'_n$ are terms, then

$$
\begin{aligned}
\mathrm{Subst}(c, x, t) &\quad \text{is} \quad c, \\
\mathrm{Subst}(x, x, t) &\quad \text{is} \quad t, \\
\mathrm{Subst}(y, x, t) &\quad \text{is} \quad y, \\
\mathrm{Subst}\big(f(t'_1, \ldots, t'_n), x, t\big) &\quad \text{is} \quad f\big(\mathrm{Subst}(t'_1, x, t), \ldots, \mathrm{Subst}(t'_n, x, t)\big). \dashv
\end{aligned}
$$

Definition 3.24 Let t be a term, and let φ be a first-order formula in which none of the variables that occur in t are bound. We define the operation $\mathrm{Subst}(\varphi, x, t)$ of substituting the term t for the variable x in

---

[4] This example and the subsequent two definitions are adapted from [17, pp. 52–54].

φ recursively on the subformulas of φ:

$$\begin{aligned}
\mathrm{Subst}(t_1 = t_2, x, t) &\quad\text{is}\quad \mathrm{Subst}(t_1, x, t) = \mathrm{Subst}(t_2, x, t), \\
\mathrm{Subst}\big(R(t_1', \ldots, t_n'), x, t\big) &\quad\text{is}\quad R\big(\mathrm{Subst}(t_1', x, t), \ldots, \mathrm{Subst}(t_n', x, t)\big), \\
\mathrm{Subst}(\neg\psi, x, t) &\quad\text{is}\quad \neg\,\mathrm{Subst}(\psi, x, t), \\
\mathrm{Subst}(\psi \circ \psi', x, t) &\quad\text{is}\quad \mathrm{Subst}(\psi, x, t) \circ \mathrm{Subst}(\psi', x, t), \\
\mathrm{Subst}\big(Qx\psi, x, t\big) &\quad\text{is}\quad Qx\varphi, \\
\mathrm{Subst}\big(Qy\psi, x, t\big) &\quad\text{is}\quad Qy\,\mathrm{Subst}(\psi, x, t).
\end{aligned}$$

Notice that y does not occur in t by hypothesis.   ⊣

We extend Definition 3.24 to all formulas by renaming bound variables as described above. More specifically, if φ is a formula in which variables *are* quantified that occur in t, we take $\mathrm{Subst}(\varphi, x, t)$ to be the same as $\mathrm{Subst}(\varphi', x, t)$, where φ' is obtained from φ by simultaneously renaming every bound variable that occurs in t in such a way as to keep distinct variables distinct.

When we substitute a term for a free variable, assignments that satisfy the original formula might not satisfy the new formula. For example, if $s(x, y) = (3, 2)$, then

$$\mathbb{N}, s \models x + y = 5.$$

If we substitute the term $2z$ for y,

$$\mathbb{N}, s \not\models x + 2z = 5.$$

because s does not assign a value to z. In contrast, if $s'(x, z) = (3, 1)$, then

$$\mathbb{N}, s' \models x + 2z = 5.$$

Now observe that $s(y) = s'(2z)$.

The previous example suggests that whenever we substitute a term for a free variable in a formula, every assignment satisfying the new formula induces a (possibly) different assignment that satisfies the original formula.

Lemma 3.25 (Substitution)   *Let φ be a first-order formula, $\mathbb{M}$ a suitable structure, and s an assignment whose domain contains $\mathrm{Free}(\varphi)$. If t is a term such that $s(t)$ is defined, then*

(a) for every term t',

$$s\big(\mathrm{Subst}(t', x, t)\big) = s\big(x/s(t)\big)(t');$$

(b) furthermore,

$$\mathbb{M}, s \models \mathrm{Subst}(\varphi, x, t) \quad iff \quad \mathbb{M}, s\big(x/s(t)\big) \models \varphi.$$

Proof (a) The proof is by induction on the complexity of t'. If t' is a constant symbol or a variable, then

$$s\big(\mathrm{Subst}(c, x, t)\big) = s(c) = s\big(x/s(t)\big)(c),$$
$$s\big(\mathrm{Subst}(x, x, t)\big) = s(t) = s\big(x/s(t)\big)(x),$$
$$s\big(\mathrm{Subst}(y, x, t)\big) = s(y) = s\big(x/s(t)\big)(y).$$

If t' is a compound term $f(t'_1, \ldots, t'_n)$, then by inductive hypothesis

$$s\big(\mathrm{Subst}(t'_i, x, t)\big) = s\big(x/s(t)\big)(t'_i).$$

Hence

$$
\begin{aligned}
s\big[\mathrm{Subst}(f(t'_1, \ldots, t'_n), x, t)\big] &= s\big[f\big(\mathrm{Subst}(t'_1, x, t), \ldots, \mathrm{Subst}(t'_n, x, t)\big)\big] \\
&= f^{\mathbb{M}}\big[s\big(x/s(t)\big)(t'_1), \ldots, s\big(x/s(t)\big)(t'_n)\big] \\
&= s\big(x/s(t)\big)\big[f(t'_1, \ldots, t'_n)\big].
\end{aligned}
$$

(b) Since renaming bound variables does not alter the meaning of a formula, we may assume that the variables in t are not quantified in φ. If φ is $t_1 = t_2$, then by part (a)

$$
\begin{aligned}
\mathbb{M}, s \models \mathrm{Subst}(t_1 = t_2, x, t) \quad &\text{iff} \quad \mathbb{M}, s \models \mathrm{Subst}(t_1, x, t) = \mathrm{Subst}(t_2, x, t) \\
&\text{iff} \quad s\big(x/s(t)\big)(t_1) = s\big(x/s(t)\big)(t_2) \\
&\text{iff} \quad \mathbb{M}, s\big(x/s(t)\big) \models t_1 = t_2.
\end{aligned}
$$

Similarly, if φ is $R(t'_1, \ldots, t'_n)$, then

$$
\begin{aligned}
\mathbb{M}, s &\models \mathrm{Subst}\big(R(t'_1, \ldots, t'_n), x, t\big) \\
&\text{iff} \quad \mathbb{M}, s \models R\big(\mathrm{Subst}(t'_1, x, t), \ldots, \mathrm{Subst}(t'_n, x, t)\big) \\
&\text{iff} \quad \big(s(x/s(t))(t'_1), \ldots, s(x/s(t))(t'_n)\big) \in R^{\mathbb{M}} \\
&\text{iff} \quad \mathbb{M}, s\big(x/s(t)\big) \models R(t'_1, \ldots, t'_n).
\end{aligned}
$$

The cases for negation, disjunction, and conjunction are straightforward applications of Theorem 3.22 to Definition 3.24. The case when φ is $Qx\psi$ depends on the fact that

$$s\big(x/s(t)\big)(x/m) = s(x/m)$$

for any $m \in M$.

If φ is $\exists y \psi$, then by inductive hypothesis

$$\mathbb{M}, s \models \mathrm{Subst}(\exists y \psi, x, t)$$

$$\begin{aligned}
&\text{iff} \quad \mathbb{M}, s \models \exists y\, \mathrm{Subst}(\psi, x, t) \\
&\text{iff} \quad \mathbb{M}, s(y/a) \models \mathrm{Subst}(\psi, x, t) & \text{(for some } a \in M) \\
&\text{iff} \quad \mathbb{M}, s(y/a)\big(x/s(y/a)(t)\big) \models \psi & \text{(for some } a \in M) \\
&\text{iff} \quad \mathbb{M}, s\big(x/s(t)\big)(y/a) \models \psi & \text{(for some } a \in M) \\
&\text{iff} \quad \mathbb{M}, s\big(x/s(t)\big) \models \exists y \psi.
\end{aligned}$$

The proof of the universal case is similar.   ⊣

3.5.2 Prenex normal form

A first-order formula is in *prenex normal form* if all of its quantifiers
appear at the front of the formula, i.e., it has the form

$$Q_1 y_1 \ldots Q_n y_n \varphi$$

where φ is quantifier free.

Theorem 3.26 (Prenex normal form) *Every first-order formula is
equivalent to a first-order formula in prenex normal form.*

Proof Let φ be a first-order formula. Without loss of generality (see
page 43) we may assume that φ is in negation normal form. As a pre-
liminary step, rename the bound variables in φ so that the resulting formula φ' has the property that every variable is quantified at most once,
and no variable occurs both free and bound. Starting with the left-most
quantifier, use Proposition 3.21 to pull each quantifier to the front of the
formula. For instance, if $\exists x \psi \vee \chi$ is a subformula of φ', then we know
$x \notin \mathrm{Free}(\chi)$ because x does not occur in χ. Hence $\exists x \psi \vee \chi \equiv \exists x(\psi \vee \chi)$.
Eventually, we obtain a formula φ'' in prenex normal form that is equiv-
alent to φ.   ⊣

For example, let φ be the formula

$$x \leq y \wedge \forall x \big[x^2 = x \vee \exists y(y < x) \big].$$

After renaming bound variables, we get

$$x \leq y \wedge \forall u \big[u^2 = u \vee \exists v(v < u) \big].$$

Pull the quantifiers to the front, one step at a time:

$$\forall u \big[x \leq y \wedge \big(u^2 = u \vee \exists v(v < u) \big) \big],$$

$$\forall u \big[x \leq y \wedge \exists v (u^2 = u \vee v < u) \big],$$

$$\forall u \exists v \big[x \leq y \wedge (u^2 = u \vee v < u) \big].$$

3.6 Satisfiability

Often when considering a first-order formula we do not have a particular structure in mind. It may suffice for a formula to be satisfied by *some* assignment in *some* model, not necessarily *this* one. At the other extreme we may require that a formula be satisfied by *every* assignment in *every* suitable model.

Definition 3.27   Let φ be a first-order formula.

- φ is *satisfiable* if there exists a suitable model $\mathbb{M}$ and an assignment s in $\mathbb{M}$ whose domain contains Free(φ) such that $\mathbb{M}, s \models \varphi$.
- φ is *valid* if for every suitable model $\mathbb{M}$ and every assignment s in $\mathbb{M}$ whose domain contains Free(φ) we have $\mathbb{M}, s \models \varphi$.   ⊣

For example, the formula $\forall x \exists y \neg R(x,y)$ is satisfiable, but not valid, while $R(x,y) \vee \neg R(x,y)$ is valid.

In the best case, it is easier to show that a formula is satisfiable than to show that it is valid because we only have to test a single model. Satisfiability and validity are really two sides of the same coin, however, because φ is valid if and only if $\neg \varphi$ is not satisfiable.

When checking the satisfiability of a sentence, we often use a process called Skolemization to eliminate existential quantifiers. First, we place the sentence in negation normal form. Then we remove the existential quantifiers one by one, replacing each of the variables bound by a given quantifier with a term involving a fresh function symbol.

Definition 3.28   Let φ be an FO_L formula in negation normal form, let U be a finite set of variables containing Free(φ), and let

$$L^* = L \cup \big\{ f_\psi : \psi \in \mathrm{Subf}_\exists(\varphi) \big\}$$

be the expansion of L obtained by adding a fresh function symbol for each existentially quantified subformula of φ. The *Skolem form* (or

Skolemization) of $\psi \in \text{Subf}(\varphi)$ with variables in U is defined recursively:

$$
\begin{aligned}
\text{Sk}_U(\psi) &\quad\text{is}\quad \psi && (\psi \text{ atomic}), \\
\text{Sk}_U(\neg\psi) &\quad\text{is}\quad \neg\,\text{Sk}_U(\psi) && (\psi \text{ atomic}), \\
\text{Sk}_U(\psi \circ \psi') &\quad\text{is}\quad \text{Sk}_U(\psi) \circ \text{Sk}_U(\psi'), \\
\text{Sk}_U(\exists x\psi) &\quad\text{is}\quad \text{Subst}\big(\text{Sk}_{U\cup\{x\}}(\psi), x, f_{\exists x\psi}(y_1, \ldots, y_n)\big), \\
\text{Sk}_U(\forall x\psi) &\quad\text{is}\quad \forall x\,\text{Sk}_{U\cup\{x\}}(\psi),
\end{aligned}
$$

where $y_1, \ldots, y_n$ enumerates the variables in U. The term

$$ f_{\exists x\psi}(y_1, \ldots, y_n) $$

is called a *Skolem term*. We abbreviate $\text{Sk}_\varnothing(\varphi)$ by $\text{Sk}(\varphi)$. If we write $\text{Sk}_U(\varphi)$ when φ is not in negation normal form, we really mean $\text{Sk}_U(\varphi')$, where φ' is the negation normal form of φ.   ⊣

For example, let φ be the sentence $\forall x \exists y\big(x < y \vee \exists z(y < z)\big)$. Then

$$
\begin{aligned}
\text{Sk}_{\{x,y,z\}}(y < z) &\quad\text{is}\quad y < z, \\
\text{Sk}_{\{x,y\}}\big(\exists z(y < z)\big) &\quad\text{is}\quad y < g(x,y), \\
\text{Sk}_{\{x,y\}}\big(x < y \vee \exists z(y < z)\big) &\quad\text{is}\quad x < y \vee y < g(x,y), \\
\text{Sk}_{\{x\}}\Big[\exists y\big(x < y \vee \exists z(y < z)\big)\Big] &\quad\text{is}\quad x < f(x) \vee f(x) < g\big(x, f(x)\big), \\
\text{Sk}(\varphi) &\quad\text{is}\quad \forall x\Big[x < f(x) \vee f(x) < g\big(x, f(x)\big)\Big].
\end{aligned}
$$

Skolemizing a first-order sentence makes explicit the dependencies between the quantified variables. Notice the difference in the Skolem forms of $\forall x \exists y R(x,y)$ and $\exists y \forall x R(x,y)$. The first is $\forall x R\big(x, f(x)\big)$, while the second is $\forall x R(x,c)$, where c is a fresh constant symbol. The Skolem forms make it clear that, in the first case, the value of the second coordinate depends on the value of the first coordinate, while in the second case it does not.

Usually, an L-structure $\mathbb{M}$ that models a sentence φ cannot model $\text{Sk}(\varphi)$ because it lacks interpretations for the fresh function symbols. We must expand $\mathbb{M}$ to an L^*-structure by specifying how we should interpret the new symbols. The interpretations of the fresh function symbols are called *Skolem functions*. There are many ways to expand a model to a larger vocabulary. Generally speaking, most of the possible expansions of $\mathbb{M}$ to the vocabulary L^* will not model $\text{Sk}(\varphi)$, but there will always be one that does.

There is another, more common, Skolemization procedure for first-order formulas in which an existentially quantified variable x is replaced by a Skolem term whose arguments are exactly those variables that are universally quantified superordinate to the existential quantifier that binds x. Both Skolemizations produce formulas that are equisatisfiable with the original formula. We defined the Skolem form of a first-order formula in terms of the Skolem forms of its subformulas in order to more easily generalize the definition in the next chapter (see Definition 4.9).

Theorem 3.29 *Let φ be an* FO_L *formula in negation normal form, and let $\mathbb{M}$ be a suitable structure. For every finite set U of variables containing* $\mathrm{Free}(\varphi)$, *there exists an expansion $\mathbb{M}^*$ of $\mathbb{M}$ to the vocabulary*

$$L^* = L \cup \left\{ f_\psi : \psi \in \mathrm{Subf}_\exists(\varphi) \right\}$$

such that for every assignment s with domain U,

$$\mathbb{M}, s \models \varphi \quad \text{iff} \quad \mathbb{M}^*, s \models \mathrm{Sk}_U(\varphi).$$

Proof We proceed by induction on the complexity of φ. If φ is atomic or negated atomic, then $\mathrm{Sk}_U(\varphi)$ is simply φ, so we can take $\mathbb{M}^* = \mathbb{M}$.

Suppose φ is $\psi_1 \vee \psi_2$. By inductive hypothesis, there exist expansions $\mathbb{M}_1^*$ and $\mathbb{M}_2^*$ of $\mathbb{M}$ to the vocabularies

$$L_1^* = L \cup \left\{ f_\chi : \chi \in \mathrm{Subf}_\exists(\psi_1) \right\},$$
$$L_2^* = L \cup \left\{ f_\chi : \chi \in \mathrm{Subf}_\exists(\psi_2) \right\},$$

respectively, such that for every assignment s with domain U,

$$\mathbb{M}, s \models \psi_1 \quad \text{iff} \quad \mathbb{M}_1^*, s \models \mathrm{Sk}_U(\psi_1),$$
$$\mathbb{M}, s \models \psi_2 \quad \text{iff} \quad \mathbb{M}_2^*, s \models \mathrm{Sk}_U(\psi_2).$$

Since the fresh function symbols in L_1^* are distinct from those in L_2^*, there is a common expansion $\mathbb{M}^*$ to the vocabulary $L^* = L_1^* \cup L_2^*$ such that

$$\mathbb{M}, s \models \psi_1 \vee \psi_2 \quad \text{iff} \quad \mathbb{M}, s \models \psi_1 \text{ or } \mathbb{M}, s \models \psi_2$$
$$\text{iff} \quad \mathbb{M}^*, s \models \mathrm{Sk}_U(\psi_1) \text{ or } \mathbb{M}^*, s \models \mathrm{Sk}_U(\psi_2)$$
$$\text{iff} \quad \mathbb{M}^*, s \models \mathrm{Sk}_U(\psi_1 \vee \psi_2).$$

The proof of the conjunction case is similar.

Suppose φ is $\exists x \psi$. By inductive hypothesis, there is an expansion $\mathbb{M}^{**}$ of $\mathbb{M}$ to the vocabulary $L^{**} = L \cup \left\{ f_\chi : \chi \in \mathrm{Subf}_\exists(\psi) \right\}$ such that for every assignment s with domain U and every $a \in M$,

$$\mathbb{M}, s(x/a) \models \psi \quad \text{iff} \quad \mathbb{M}^{**}, s(x/a) \models \mathrm{Sk}_{U \cup \{x\}}(\psi).$$

Let g be a function mapping assignments to individuals with the property that for every assignment s with domain U, if there exists an $a \in M$ such that $M, s(x/a) \models \psi$, then $M, s(x/g(s)) \models \psi$. Expand M^{**} to M^* by defining

$$f^{M^*}_{\exists x \psi}(s(y_1), \ldots, s(y_n)) = g(s),$$

where $y_1, \ldots, y_n$ enumerates U. Then by the substitution lemma,

$M, s \models \exists x \psi$

    iff $M, s(x/g(s)) \models \psi$

    iff $M^{**}, s(x/g(s)) \models \mathrm{Sk}_{U \cup \{x\}}(\psi)$

    iff $M^*, s\left(x \big/ f^{M^*}_{\exists x \psi}(s(y_1), \ldots, s(y_n))\right) \models \mathrm{Sk}_{U \cup \{x\}}(\psi)$

    iff $M^*, s \models \mathrm{Subst}\left(\mathrm{Sk}_{U \cup \{x\}}(\psi), x, f_{\exists x \psi}(y_1, \ldots, y_n)\right)$

    iff $M^*, s \models \mathrm{Sk}_U(\exists x \psi).$

Finally, suppose ψ is $\forall x \psi$. By inductive hypothesis, there is an expansion M^{**} of M to the vocabulary $L^{**} = L \cup \{ f_\chi : \chi \in \mathrm{Subf}_\exists(\psi) \}$ such that for every assignment s with domain U and every $a \in M$,

$$M, s(x/a) \models \psi \quad \text{iff} \quad M^{**}, s(x/a) \models \mathrm{Sk}_{U \cup \{x\}}(\psi).$$

Since the vocabulary L^* is the same as L^{**}, we can take $M^* = M^{**}$. Then

$M, s \models \forall x \psi$ iff $M, s(x/a) \models \psi$ (for all $a \in M$)

           iff $M^*, s(x/a) \models \mathrm{Sk}_{U \cup \{x\}}(\psi)$ (for all $a \in M$)

           iff $M^*, s \models \forall x\, \mathrm{Sk}_{U \cup \{x\}}(\psi)$

           iff $M^*, s \models \mathrm{Sk}_U(\forall x \psi).$ ⊣

When discussing whether M, s satisfies an existential formula $\exists x \varphi$, an element $a \in M$ such that $M, s(x/a) \models \varphi$ is called a *witness* to $\exists x \varphi$. Thus one can say that Skolem functions point out witnesses to existential formulas. A universal formula $\forall x \varphi$ is not satisfied by M, s if and only if M, s satisfies $\exists x(\neg \varphi)$. An element $a \in M$ such that $M, s(x/a) \models \neg\varphi$ is called a *Kreisel counterexample* to $\forall x \varphi$. For example, a Kreisel counterexample to the sentence

$$\forall x \exists y R(x, y)$$

is an element a such that for all b we have $(a, b) \notin R^M$.

Skolem functions and Kreisel counterexamples will play an important role in the next chapter.

4

Independence-friendly logic

In the last chapter, we studied first-order logic from the game-theoretic perspective. In particular, we defined the semantic game for a first-order formula as an extensive game with perfect information. It is natural to extend first-order logic by considering semantic games with imperfect information. Hintikka and Sandu named this extension *independence-friendly logic* because in a semantic game with imperfect information the choices made by the players may be independent of prior moves [30].

In this chapter, we define the syntax of independence-friendly (IF) logic and three different semantics. The first semantics is game-theoretic; the second is based on Skolem functions and Kreisel counterexamples. The third semantics, due to Hodges [32, 33], is modeled after Tarski's compositional semantics for first-order logic. At the end of the chapter, we prove that all three semantics are equivalent.[1]

4.1 Syntax

Definition 4.1 Let L be a vocabulary. L-terms are defined as for first-order logic. The *independence-friendly language* IF_L is generated from L according to the following rules:

- If t_1 and t_2 are L-terms, then $(t_1 = t_2) \in \mathrm{IF}_L$ and $\neg(t_1 = t_2) \in \mathrm{IF}_L$.
- If R is an n-ary relation symbol in L and $t_1, \ldots, t_n$ are L-terms, then $R(t_1, \ldots, t_n) \in \mathrm{IF}_L$ and $\neg R(t_1, \ldots, t_n) \in \mathrm{IF}_L$.
- If $\varphi, \varphi' \in \mathrm{IF}_L$, then $(\varphi \vee \varphi') \in \mathrm{IF}_L$ and $(\varphi \wedge \varphi') \in \mathrm{IF}_L$.

---

[1] The authors wish to acknowledge the contribution to the study of IF logic made by Dechesne's PhD thesis [14]. Many of our notational and conceptual choices were influenced by her work.

- If $\varphi \in \mathrm{IF}_L$, x is a variable, and W is a finite set of variables,
  then $(\exists x/W)\varphi \in \mathrm{IF}_L$ and $(\forall x/W)\varphi \in \mathrm{IF}_L$.

The elements of IF_L are called IF_L *formulas*. When the vocabulary is irrelevant or clear from context we will not mention it explicitly. An IF *formula* is an IF_L formula for some vocabulary L.   ⊣

To simplify the presentation, we only allow the negation symbol ¬ to appear in front of atomic formulas. We will see in Chapter 5 that this restriction is not essential; it simply allows us to assume that Eloise is always the verifier. Formulas of the form $(t_1 = t_2)$, $\neg(t_1 = t_2)$, $R(t_1, \ldots, t_n)$, or $\neg R(t_1, \ldots, t_n)$ are called *literals*. When φ is a literal, we abuse our notation slightly by writing $\neg\varphi$ for the dual of φ. When φ is a literal and φ' is any IF formula we use $\varphi \to \varphi'$ as an abbreviation for $\neg\varphi \lor \varphi'$.

The finite set of variables W in $(\exists x/W)$ and $(\forall x/W)$ is called a *slash set*. A slash set indicates from which variables a quantifier is independent. We write $\exists x$ and $\forall x$ for $(\exists x/\varnothing)$ and $(\forall x/\varnothing)$, respectively. Thus every first-order formula is a shorthand for an IF formula.

The subformula tree of an IF formula is like the subformula tree of a first-order formula except that we do not take atomic formulas to be subformulas of their negations. Therefore a leaf of the tree may be any literal.

Definition 4.2   Let φ be an IF formula. The *subformulas* of φ are defined recursively:

$$\mathrm{Subf}(\psi) = \{\psi\} \quad (\psi \text{ literal}),$$
$$\mathrm{Subf}(\psi \circ \psi') = \{\psi \circ \psi'\} \cup \mathrm{Subf}(\psi) \cup \mathrm{Subf}(\psi'),$$
$$\mathrm{Subf}((Qx/W)\psi) = \{(Qx/W)\psi\} \cup \mathrm{Subf}(\psi).$$

As with first-order formulas, we distinguish between multiple occurrences of the same subformula. The set of literal subformulas of φ is denoted $\mathrm{Lit}(\varphi)$.   ⊣

The set of *free variables* of an IF formula φ is defined as for first-order formulas (Definition 3.5) except that the clause for quantifiers is replaced by:

$$\mathrm{Free}((Qx/W)\varphi) = \big(\mathrm{Free}(\varphi) - \{x\}\big) \cup W.$$

An occurrence of a variable x is *bound* by the innermost quantifier

(Qx/W) in whose scope it occurs. The set of bound variables of φ is denoted Bound(φ). For example, in the formula

$$\forall x\,(\exists y/\{x\})\,R(x,y) \wedge \forall y\,(\exists z/\{x,y\})\,R(y,z),$$

the variables y and z are bound, while x is both free and bound. An IF formula with no free variables is called an IF *sentence*.

Formulas such as $\forall y\,(\exists z/\{x,y\})\,R(y,z)$ in which a variable occurs in a slash set but nowhere else may strike the reader as a bit odd. What does it mean for the quantifier $(\exists z/\{x,y\})$ to be independent of a variable that is not quantified superordinate to $(\exists z/\{x,y\})$? In a moment, we will see that such formulas *do* make sense, but in the meantime we reassure ourselves by defining a restricted class of IF formulas for which such questions do not arise.

Definition 4.3   An IF formula is *regular* if it satisfies the following conditions:

(a) If (Qy/W) is a quantifier such that $x \in W$, then it is subordinate to a quantifier of the form $(\exists x/V)$ or $(\forall x/V)$.
(b) No quantifier of the form (Qx/W) is subordinate to a quantifier of the form $(\exists x/V)$ or $(\forall x/V)$.   ⊣

The first condition ensures that a quantifier is only independent of superordinate quantifiers, while the second forbids double quantification. For example, the formula $\forall x\,(\exists y/\{x\})\,x = y$ is regular, while the formulas

$$(\exists y/\{x\})x = y \quad \text{and} \quad \exists x(\varphi \wedge \exists x\psi)$$

are not. The impact of the two constraints can be better grasped if we keep in mind that IF formulas will be interpreted by semantic games. Each quantifier (Qx/V) prompts the appropriate player to choose an individual from the universe to be the value of x. However, the player's choice must not depend on the values of the variables in V. The first constraint means that in the semantic game for a regular IF formula, if the choice of y is independent of x, then the value of x must have been specified at some earlier stage of the game, while the second constraint implies that in a given play of the game, the value of a variable is only specified once. Observe that in a regular IF formula it is not possible to have a quantifier (Qx/W) such that $x \in W$ because by the first condition we would need to have another quantifier of the form $(\exists x/V)$ or $(\forall x/V)$ superordinate to (Qx/W), which is forbidden by the second condition.

4.2 Game-theoretic semantics

Now that we have defined the syntax of IF logic, we are ready to say
what IF formulas mean. In the last chapter, we observed that a first-
order formula could be interpreted as specifying a game with perfect
information. In this section, we interpret IF formulas as specifying a
game with imperfect information.

We will define the semantic game for an IF formula as an extensive
game with imperfect information by restricting the players' access to the
current assignment. That is, a player may be forced to choose an action
without knowing the current assignment in its entirety.

Definition 4.4   Two assignments, s and s', such that $W \subseteq \mathrm{dom}(s) =
\mathrm{dom}(s')$ are *equivalent modulo W* (or *W-equivalent*), denoted $s \approx_W s'$,
if for every variable $x \in \mathrm{dom}(s) - W$ we have $s(x) = s'(x)$. $\dashv$

Definition 4.5   Let φ be an IF formula, $\mathbb{M}$ a suitable structure, and
s an assignment whose domain contains Free(φ). The *semantic game*
$G(\mathbb{M}, s, \varphi)$ is a win-lose extensive game with imperfect information:

- There are two players, Eloise ($\exists$) and Abelard ($\forall$).
- The set of histories is $H = \bigcup \{\, H_\psi : \psi \in \mathrm{Subf}(\varphi) \,\}$, where H_ψ is
  defined recursively:

  - $H_\varphi = \{(s, \varphi)\}$,
  - if ψ is $\chi_1 \circ \chi_2$, then $H_{\chi_i} = \{\, h^\frown \chi_i : h \in H_{\chi_1 \circ \chi_2} \,\}$,
  - if ψ is $(Qx/W)\chi$, then $H_\chi = \{\, h^\frown(x, a) : h \in H_{(Qx/W)\chi}, a \in M \,\}$.

  Observe that (s, φ) is the unique initial history. The assignment s is
  called the *initial assignment*. Every history h' induces an assignment
  $s_{h'}$ extending and/or modifying the initial assignment:

$$
s_{h'} = \begin{cases}
s & \text{if } h' = (s, \varphi), \\
s_h & \text{if } h' = h^\frown \chi, \\
s_h(x/a) & \text{if } h' = h^\frown(x, a).
\end{cases}
$$

- Once play reaches a literal, the game ends:

$$
Z = \bigcup \{\, H_\chi : \chi \in \mathrm{Lit}(\varphi) \,\}.
$$

- Disjunctions and existential quantifiers are decision points for Eloise,
  while conjunctions and universal quantifiers are decision points for

Abelard:

$$P(h) = \begin{cases} \exists & \text{if } h \in H_{\chi \vee \chi'} \text{ or } h \in H_{(\exists x/W)\chi}, \\ \forall & \text{if } h \in H_{\chi \wedge \chi'} \text{ or } h \in H_{(\forall x/W)\chi}. \end{cases}$$

- The indistinguishability relations $\sim_\exists$ and $\sim_\forall$ are defined as follows. For all $h, h' \in H_{\chi \vee \chi'}$ we have $h \sim_\exists h'$ if and only if $s_h = s_{h'}$. For all $h, h' \in H_{(\exists x/W)\chi}$,

$$h \sim_\exists h' \quad \text{iff} \quad s_h \approx_W s_{h'}.$$

Similarly, for all $h, h' \in H_{\chi \wedge \chi'}$ we have $h \sim_\forall h'$ if and only if $s_h = s_{h'}$, and for all $h, h' \in H_{(\forall x/W)\chi}$,

$$h \sim_\forall h' \quad \text{iff} \quad s_h \approx_W s_{h'}.$$

- Eloise wins if the literal χ reached at the end of play is satisfied by the current assignment; Abelard wins if it is not:

$$u(h) = \begin{cases} \exists & \text{if } \mathbb{M}, s_h \models \chi, \\ \forall & \text{if } \mathbb{M}, s_h \not\models \chi. \end{cases} \qquad \dashv$$

The equivalence relations $\sim_\exists$ and $\sim_\forall$ specify exactly how much information the players have at their disposal at a given decision point. For example, in the position of the game corresponding to $(\exists x/W)\psi$, Eloise cannot distinguish histories whose induced assignments agree on the variables outside of W but disagree on variables in W. Hence, Eloise must choose the value of x without having access to the values of the variables in W. Likewise for Abelard in the position corresponding to $(\forall x/W)\psi$. Note that if h is a history whose current subformula is $(\exists x/W)\psi$ or $(\forall x/W)\psi$ then

$$A(h) = \big\{ (x, a) : a \in M \big\}.$$

That is, limiting the information available to a player does not affect his or her ability to perform any particular action. It simply prevents the player from employing certain strategies.

Since the semantic game for an IF formula is a game with imperfect information, we cannot rely on the Gale-Stewart theorem to guarantee that one of the two players will have a winning strategy. Therefore, we must be careful about how we define truth and satisfaction in IF logic.

Definition 4.6 Let φ be an IF formula, $\mathbb{M}$ a suitable structure, and s an assignment whose domain contains Free(φ).

$\mathbb{M}, s \models^+_{\text{GTS}} \varphi$ iff   Eloise has a winning strategy for $G(\mathbb{M}, s, \varphi)$.

$\mathbb{M}, s \models^-_{\text{GTS}} \varphi$ iff   Abelard has a winning strategy for $G(\mathbb{M}, s, \varphi)$.

In the first case we say $\mathbb{M}, s$ *satisfies* φ, and in the second case we say $\mathbb{M}, s$ *dissatisfies* φ.

An IF sentence φ is *true* in $\mathbb{M}$, denoted $\mathbb{M} \models^+_{\text{GTS}} \varphi$, if it is satisfied by the empty assignment. It is *false*, denoted $\mathbb{M} \models^-_{\text{GTS}} \varphi$, if it is dissatisfied by the empty assignment.   ⊣

Example 4.7 In the game Matching Pennies there are two players. Each player has a coin that he or she secretly turns to heads or tails. The coins are revealed simultaneously. The first player wins if the coins are both heads or both tails; the second player wins if they differ.

We can express the game Matching Pennies using the IF sentence

$$\forall x (\exists y / \{x\}) x = y$$

interpreted in the two-element structure $\mathbb{M} = \{a, b\}$. We wish to show that neither Eloise nor Abelard has a winning strategy, assuming the initial assignment is empty. Call the original sentence φ_{MP}, and let ψ be the subformula

$$(\exists y / \{x\}) x = y.$$

Then $H_{\varphi_{\text{MP}}}$ includes only the initial history $(\varnothing, \varphi_{\text{MP}})$, while H_ψ includes two histories: $h_a = (\varnothing, \varphi_{\text{MP}}, (x, a))$ and $h_b = (\varnothing, \varphi_{\text{MP}}, (x, b))$. Let σ be a strategy for Eloise. Since $h_a \sim_\exists h_b$ she must choose the same value for y in both cases:

$$\sigma(h_a) = (y, c) = \sigma(h_b).$$

Hence there are two maximal plays,

$$(\varnothing, \varphi_{\text{MP}}, (x, a), (y, c)) \quad \text{and} \quad (\varnothing, \varphi_{\text{MP}}, (x, b), (y, c)),$$

of which Eloise wins exactly one.

Now let τ be a strategy for Abelard such that $\tau(\varnothing, \varphi_{\text{MP}}) = (x, c)$. Then τ is a winning strategy if and only if Abelard wins both maximal plays $(\varnothing, \varphi_{\text{MP}}, (x, c), (y, a))$ and $(\varnothing, \varphi_{\text{MP}}, (x, c), (y, b))$ which is again impossible (see Figure 4.1).   ⊣

According to our definition the sentence $\forall x (\exists y / \{x\}) x = y$ is neither true nor false in any structure with at least two elements. Those who have

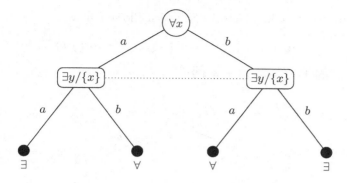

Figure 4.1 The semantic game for $\forall x (\exists y / \{x\}) x = y$ in the structure $\mathbb{M} = \{a, b\}$

studied games with imperfect information will be unsurprised, but logicians may find the failure of the principle of bivalence unsettling. There are other surprises in store for logicians. Adding a superfluous quantifier does not affect the truth value of a first-order sentence. For example $\forall x \exists y R(x, y)$ and $\forall x \exists y \exists y R(x, y)$ are equivalent. In contrast, adding extra quantifiers *can* affect the truth-value of an IF sentence.

Example 4.8   We add one dummy quantifier $\exists y$ to the sentence in Example 4.7 to get the irregular IF sentence

$$\forall x \exists y (\exists y / \{x\}) x = y,$$

which we interpret in the two-element structure $\mathbb{M} = \{a, b\}$. Surprisingly, Eloise has a winning strategy. For convenience, let ψ be the subformula

$$\exists y (\exists y / \{x\}) x = y,$$

and let χ be the subformula $(\exists y / \{x\}) x = y$. Then H_ψ is as before, while H_χ consists of four histories:

$$h_{aa} = \big(\varnothing, \varphi, (x, a), (y, a)\big), \qquad h_{ab} = \big(\varnothing, \varphi, (x, a), (y, b)\big),$$
$$h_{ba} = \big(\varnothing, \varphi, (x, b), (y, a)\big), \qquad h_{bb} = \big(\varnothing, \varphi, (x, b), (y, b)\big).$$

Observe that $h_{aa} \sim_\exists h_{ba}$ and $h_{ab} \sim_\exists h_{bb}$. Therefore all Eloise's strategies must satisfy $\sigma(h_{aa}) = \sigma(h_{ba})$ and $\sigma(h_{ab}) = \sigma(h_{bb})$. Here is a winning strategy:

$$\sigma(h_a) = (y, a) \quad \text{and} \quad \sigma(h_{aa}) = \sigma(h_{ba}) = (y, a),$$
$$\sigma(h_b) = (y, b) \quad \text{and} \quad \sigma(h_{ab}) = \sigma(h_{bb}) = (y, b).$$

There are two terminal histories in which Eloise follows σ,

$$(\varnothing, \varphi, (x, a), (y, a), (y, a)) \quad \text{and} \quad (\varnothing, \varphi, (x, b), (y, b), (y, b)),$$

and she wins both (see Figure 4.2). ⊣

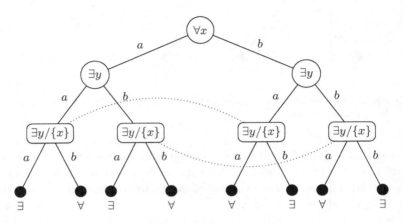

Figure 4.2 The semantic game for $\forall x \exists y (\exists y/\{x\}) x = y$ in $\mathbb{M} = \{a, b\}$

By allowing Eloise to choose the value of y twice, we enable her to copy the value of x to a location that she is allowed to see when making her second choice. There are two ways to block Eloise's winning strategy. We can prevent her from seeing the value of x both times she chooses the value of y,

$$\forall x (\exists y/\{x\})(\exists y/\{x\}) x = y,$$

or we can prevent her from seeing the value of y when making her second choice,

$$\forall x \exists y (\exists y/\{x, y\}) x = y.$$

The fact that Eloise assigns a value to the same variable twice is inconsequential. She has a similar winning strategy for $\forall x \exists z (\exists y/\{x\}) x = y$.

Such phenomena are common in games of imperfect information. In bridge, skilled partners can communicate to each other about their hands using only the cards they play. Playing according to a predetermined convention in order to circumvent informational restrictions is called *signaling*. The possibility of signaling in IF logic was first observed by Hodges [32].

4.3 Skolem semantics

In this section we describe another interpretation of IF formulas based on Skolem functions. To this end, we generalize the Skolemization procedure for first-order logic to IF sentences.

Definition 4.9 Let φ be an IF_L formula, let U be a finite set of variables containing $\text{Free}(\varphi)$, and let

$$L^* = L \cup \{\, f_\psi : \psi \in \text{Subf}_\exists(\varphi) \,\}$$

be the expansion of L obtained by adding a fresh function symbol for every existentially quantified subformula of φ. The *Skolem form* (or *Skolemization*) of $\psi \in \text{Subf}(\varphi)$ with variables in U is defined recursively:

$$
\begin{aligned}
\text{Sk}_U(\psi) &\ \text{is} \ \ \psi && (\psi \ \text{literal}), \\
\text{Sk}_U(\psi \circ \psi') &\ \text{is} \ \ \text{Sk}_U(\psi) \circ \text{Sk}_U(\psi'), \\
\text{Sk}_U\big((\exists x/W)\psi\big) &\ \text{is} \ \ \text{Subst}\big(\text{Sk}_{U \cup \{x\}}(\psi), x, f_{(\exists x/W)\psi}(y_1, \ldots, y_n)\big), \\
\text{Sk}_U\big((\forall x/W)\psi\big) &\ \text{is} \ \ \forall x \, \text{Sk}_{U \cup \{x\}}(\psi),
\end{aligned}
$$

where $y_1, \ldots, y_n$ enumerates the variables in $U - W$. Observe that at each stage $\text{Sk}_U(\psi)$ is a first-order formula. We abbreviate $\text{Sk}_\varnothing(\varphi)$ by $\text{Sk}(\varphi)$. ⊣

The reader should notice that in the context of $\text{Sk}_U\big((\exists x/W)\psi\big)$, the variable x may belong to U. If it does, then x will be an argument of the Skolem term $f_{(\exists x/W)}(y_1, \ldots, y_n)$ unless x also belongs to W. Also notice that $\text{Sk}_U\big((\forall x/W)\psi\big) = \text{Sk}_{U \cup \{x\}}\big((\forall x/W)\psi\big)$.

Example 4.10 Examine the Skolem form of the Matching Pennies sentence $\forall x(\exists y/\{x\})x = y$. Proceeding inside-out,

$$
\begin{aligned}
\text{Sk}_{\{x,y\}}(x = y) &\ \text{is} \ \ x = y, \\
\text{Sk}_{\{x\}}\big[(\exists y/\{x\})x = y\big] &\ \text{is} \ \ x = c, \\
\text{Sk}\big[\forall x(\exists y/\{x\})x = y\big] &\ \text{is} \ \ \forall x(x = c),
\end{aligned}
$$

where c is a fresh constant symbol. ⊣

Example 4.11 Now consider the Skolem form of the Matching Pennies sentence augmented with a dummy quantifier, $\forall x \exists y(\exists y/\{x\})x = y$:

$$\mathrm{Sk}_{\{x,y\}}(x = y) \quad \text{is} \quad x = y,$$
$$\mathrm{Sk}_{\{x,y\}}\Big[(\exists y/\{x\})x = y\Big] \quad \text{is} \quad x = g(y),$$
$$\mathrm{Sk}_{\{x\}}\Big[\exists y(\exists y/\{x\})x = y\Big] \quad \text{is} \quad x = g(f(x)),$$
$$\mathrm{Sk}\Big[\forall x \exists y(\exists y/\{x\})x = y\Big] \quad \text{is} \quad \forall x\Big[x = g(f(x))\Big].$$

If we use a distinct variable for the dummy quantifier, as in

$$\forall x \exists z(\exists y/\{x\})x = y,$$

the Skolemization procedure yields the same Skolem form:

$$\mathrm{Sk}_{\{x,y,z\}}(x = y) \quad \text{is} \quad x = y,$$
$$\mathrm{Sk}_{\{x,z\}}\Big[(\exists y/\{x\})x = y\Big] \quad \text{is} \quad x = g(z),$$
$$\mathrm{Sk}_{\{x\}}\Big[\exists z(\exists y/\{x\})x = y\Big] \quad \text{is} \quad x = g(f(x)),$$
$$\mathrm{Sk}\Big[\forall x \exists z(\exists y/\{x\})x = y\Big] \quad \text{is} \quad \forall x\Big[x = g(f(x))\Big]. \qquad \dashv$$

We are now ready to present a second way of interpreting IF formulas, called *Skolem semantics*.

Definition 4.12 Let φ be an IF_L formula, $\mathbb{M}$ a suitable structure, and s an assignment whose domain contains $\mathrm{Free}(\varphi)$. Define

$$\mathbb{M}, s \models^+_{\mathrm{Sk}} \varphi \quad \text{iff} \quad \mathbb{M}^*, s \models \mathrm{Sk}_{\mathrm{dom}(s)}(\varphi)$$

for some expansion $\mathbb{M}^*$ of $\mathbb{M}$ to the vocabulary

$$L^* = L \cup \{ f_\psi : \psi \in \mathrm{Subf}_\exists(\varphi) \}. \qquad \dashv$$

When evaluating an IF formula under Skolem semantics, we implicitly assume that every variable that has been assigned a value is "present" in the formula. Thus the Skolemization of an IF formula depends on the assignment used to evaluate it. For example, suppose s and s' are assignments such that $\mathrm{dom}(s) = \{u, v\}$ and $\mathrm{dom}(s') = \{u, v, w\}$. Then

$$\mathbb{M}, s \models^+_{\mathrm{Sk}} (\exists x/\{u\})P(x) \quad \text{iff} \quad \mathbb{M}^*, s \models P(f(v))$$

for some expansion $\mathbb{M}^*$ of $\mathbb{M}$, while

$$\mathbb{M}, s' \models^+_{\mathrm{Sk}} (\exists x/\{u\})P(x) \quad \text{iff} \quad \mathbb{M}^{**}, s' \models P(g(v, w))$$

for some expansion $\mathbb{M}^{**}$ of $\mathbb{M}$.

The first thing to check is that Skolem semantics agrees with the game-theoretic semantics defined in the previous section.

Theorem 4.13 *Let φ be an IF_L formula, $\mathbb{M}$ a suitable structure, and s an assignment whose domain contains $\mathrm{Free}(\varphi)$. Then*

$$\mathbb{M}, s \models^+_{\mathrm{GTS}} \varphi \quad \textit{iff} \quad \mathbb{M}, s \models^+_{\mathrm{Sk}} \varphi.$$

Proof Suppose Eloise has a winning strategy σ for $G(\mathbb{M}, s, \varphi)$. Let $\mathbb{M}^*$ be an expansion of $\mathbb{M}$ to the vocabulary

$$L^* = L \cup \{ f_\psi : \psi \in \mathrm{Subf}_\exists(\varphi) \}$$

such that for every existential subformula $(\exists x/W)\psi'$ of φ and every history $h \in H_{(\exists x/W)\psi'}$,

$$f^{\mathbb{M}^*}_{(\exists x/W)\psi'}(s_h(y_1), \ldots, s_h(y_n)) = a,$$

where $y_1, \ldots, y_n$ enumerates $\mathrm{dom}(s_h) - W$, and $\sigma(h) = (x, a)$. To show the function is well defined, suppose $h, h' \in H_{(\exists x/W)\psi'}$ are two histories such that

$$\sigma(h) = (x, a) \neq (x, a') = \sigma(h').$$

Then $s_h \not\approx_W s_{h'}$ which means $s_h(y_i) \neq s_{h'}(y_i)$ for some $y_i \in \{y_1, \ldots, y_n\}$.

It suffices to show that for all $\psi \in \mathrm{Subf}(\varphi)$, if Eloise follows σ in $h \in H_\psi$, then $\mathbb{M}^*, s_h \models \mathrm{Sk}_{\mathrm{dom}(s_h)}(\psi)$ (because Eloise does follow σ in $(s, \varphi) \in H_\varphi$). If ψ is a literal, and Eloise follows σ in $h \in H_\psi$, then $\mathbb{M}, s_h \models \psi$ because σ is a winning strategy. Hence $\mathbb{M}^*, s_h \models \mathrm{Sk}_{\mathrm{dom}(s_h)}(\psi)$.

Suppose ψ is $\psi_1 \vee \psi_2$. If Eloise follows σ in $h \in H_{\psi_1 \vee \psi_2}$, and $\sigma(h) = \psi_i$, then Eloise follows σ in $h' = h^\frown \psi_i$. By inductive hypothesis

$$\mathbb{M}^*, s_{h'} \models \mathrm{Sk}_{\mathrm{dom}(s_{h'})}(\psi_i),$$

whence

$$\mathbb{M}^*, s_{h'} \models \mathrm{Sk}_{\mathrm{dom}(s_{h'})}(\psi_1) \vee \mathrm{Sk}_{\mathrm{dom}(s_{h'})}(\psi_2).$$

Since $s_h = s_{h'}$, it follows that $\mathbb{M}^*, s_h \models \mathrm{Sk}_{\mathrm{dom}(s_h)}(\psi_1 \vee \psi_2)$.

Suppose ψ is $\psi_1 \wedge \psi_2$. If Eloise follows σ in $h \in H_{\psi_1 \wedge \psi_2}$, then she follows σ in both $h_1 = h^\frown \psi_1$ and $h_2 = h^\frown \psi_2$. By inductive hypothesis $\mathbb{M}^*, s_{h_1} \models \mathrm{Sk}_{\mathrm{dom}(s_{h_1})}(\psi_1)$ and $\mathbb{M}^*, s_{h_2} \models \mathrm{Sk}_{\mathrm{dom}(s_{h_2})}(\psi_2)$, whence

$$\mathbb{M}^*, s_h \models \mathrm{Sk}_{\mathrm{dom}(s_h)}(\psi_1) \wedge \mathrm{Sk}_{\mathrm{dom}(s_h)}(\psi_2).$$

It follows that $\mathbb{M}^*, s_h \models \mathrm{Sk}_{\mathrm{dom}(s_h)}(\psi_1 \wedge \psi_2)$.

Suppose ψ is $(\exists x/W)\psi'$. If Eloise follows σ in $h \in H_{(\exists x/W)\psi'}$, and $\sigma(h) = (x, a)$, then Eloise follows σ in $h' = h^\frown(x, a)$. By inductive hypothesis $\mathbb{M}^*, s_{h'} \models \mathrm{Sk}_{\mathrm{dom}(s_{h'})}(\psi')$, which is to say

$$\mathbb{M}^*, s_h(x/a) \models \mathrm{Sk}_{\mathrm{dom}(s_h(x/a))}(\psi').$$

By construction $f_{(\exists x/W)\psi'}^{\mathbb{M}^*}(s_h(y_1), \ldots, s_h(y_n)) = a$, where $y_1, \ldots, y_n$ enumerates $\mathrm{dom}(s_h) - W$, so an application of the substitution lemma (Lemma 3.25) yields

$$\mathbb{M}^*, s_h \models \mathrm{Subst}\big(\mathrm{Sk}_{\mathrm{dom}(s_h(x/a))}(\psi'), x, f_{(\exists x/W)\psi'}(y_1, \ldots, y_n)\big).$$

Hence $\mathbb{M}^*, s_h \models \mathrm{Sk}_{\mathrm{dom}(s_h)}\big((\exists x/W)\psi'\big)$.

Suppose ψ is $(\forall x/W)\psi'$. If Eloise follows σ in $h \in H_{(\forall x/W)\psi'}$, then she follows σ in every $h_a = h^\frown(x, a) \in H_{\psi'}$. By inductive hypothesis $\mathbb{M}^*, s_{h_a} \models \mathrm{Sk}_{\mathrm{dom}(s_{h_a})}(\psi')$. Since $s_{h_a} = s_h(x/a)$, it follows that

$$\mathbb{M}^*, s_h \models \forall x\, \mathrm{Sk}_{\mathrm{dom}(s_h) \cup \{x\}}(\psi'),$$

which implies $\mathbb{M}^*, s_h \models \mathrm{Sk}_{\mathrm{dom}(s_h)}\big((\forall x/W)\psi'\big)$.

Conversely, suppose there is an expansion $\mathbb{M}^*$ of $\mathbb{M}$ such that

$$\mathbb{M}^*, s \models \mathrm{Sk}_{\mathrm{dom}(s)}(\varphi).$$

Let σ be the strategy for Eloise defined as follows. If $h \in H_{\psi_1 \vee \psi_2}$, then

$$\sigma(h) = \begin{cases} \psi_1 & \text{if } \mathbb{M}^*, s_h \models \mathrm{Sk}_{\mathrm{dom}(s_h)}(\psi_1), \\ \psi_2 & \text{otherwise.} \end{cases}$$

If $h \in H_{(\exists x/W)\psi'}$, then

$$\sigma(h) = \Big(x, f_{(\exists x/W)\psi'}^{\mathbb{M}^*}\big(s_h(y_1), \ldots, s_h(y_n)\big)\Big),$$

where $y_1, \ldots, y_n$ enumerates $\mathrm{dom}(s_h) - W$.

We show by induction on the length of h that if Eloise follows σ in $h \in H_\psi$, then $\mathbb{M}^*, s_h \models \mathrm{Sk}_{\mathrm{dom}(s_h)}(\psi)$. The basis step follows from the original supposition. For the inductive step, suppose Eloise follows σ in

$$h' = (s, \varphi, a_1, \ldots, a_m, a_{m+1}).$$

Then she certainly follows σ in $h = (s, \varphi, a_1, \ldots, a_m)$.

Suppose $h \in H_{\psi_1 \vee \psi_2}$ and $a_{m+1} = \psi_i$. Then by inductive hypothesis $\mathbb{M}^*, s_h \models \mathrm{Sk}_{\mathrm{dom}(s_h)}(\psi_1 \vee \psi_2)$, so by construction $\mathbb{M}^*, s_{h'} \models \mathrm{Sk}_{\mathrm{dom}(s_{h'})}(\psi_i)$.

Suppose $h \in H_{\psi_1 \wedge \psi_2}$. Then by inductive hypothesis

$$\mathbb{M}^*, s_h \models \mathrm{Sk}_{\mathrm{dom}(s_h)}(\psi_1 \wedge \psi_2),$$

from which it follows that $\mathbb{M}^*, s_{h'} \models \mathrm{Sk}_{\mathrm{dom}(s_{h'})}(\psi_i)$.

Suppose $h \in H_{(\exists x/W)\psi'}$ and $a_{m+1} = (x, a)$. By inductive hypothesis $\mathbb{M}^*, s_h \models \mathrm{Sk}_{\mathrm{dom}(s_h)}((\exists x/W)\psi')$, which is to say

$$\mathbb{M}^*, s_h \models \mathrm{Subst}\big(\mathrm{Sk}_{\mathrm{dom}(s_h) \cup \{x\}}(\psi'), x, f_{(\exists x/W)\psi}(y_1, \dots, y_n)\big),$$

where $y_1, \dots, y_n$ enumerates $\mathrm{dom}(s_h) - W$. By the substitution lemma,

$$\mathbb{M}^*, s_h(x/a) \models \mathrm{Sk}_{\mathrm{dom}(s_h) \cup \{x\}}(\psi'),$$

which implies $\mathbb{M}^*, s_{h'} \models \mathrm{Sk}_{\mathrm{dom}(s_{h'})}(\psi')$.

Suppose $h \in H_{(\forall x/W)\psi'}$. Then by inductive hypothesis

$$\mathbb{M}^*, s_h \models \mathrm{Sk}_{\mathrm{dom}(s_h)}((\forall x/W)\psi'),$$

which is to say $\mathbb{M}^*, s_h \models \forall x\, \mathrm{Sk}_{\mathrm{dom}(s_h) \cup \{x\}}(\psi')$. It follows that

$$\mathbb{M}^*, s_{h'} \models \mathrm{Sk}_{\mathrm{dom}(s_{h'})}(\psi').$$

Finally, observe that if Eloise follows σ in a terminal history $h \in H_\chi$, then $\mathbb{M}^*, s_h \models \mathrm{Sk}_{\mathrm{dom}(s_h)}(\chi)$. It follows that $\mathbb{M}, s_h \models \chi$, so Eloise wins h. Therefore σ is a winning strategy for Eloise. ⊣

We can use Skolem semantics to give examples of IF sentences that express concepts that are not definable in ordinary first-order logic. Our first example is adapted from [10, Example 1.4].

Example 4.14 Let φ_∞ be the IF sentence

$$\exists w \forall x (\exists y/\{w\})(\exists z/\{w, x\})[z = x \wedge y \neq w],$$

and let ψ be the subformula $[z = x \wedge y \neq w]$. The Skolem form of φ_∞ is obtained in the following stages:

$$
\begin{aligned}
\mathrm{Sk}_{\{w,x,y,z\}}(\psi) \quad &\text{is} \quad z = x \wedge y \neq w, \\
\mathrm{Sk}_{\{w,x,y\}}\big[(\exists z/\{w,x\})\,\psi\big] \quad &\text{is} \quad g(y) = x \wedge y \neq w, \\
\mathrm{Sk}_{\{w,x\}}\big[(\exists y/\{w\})(\exists z/\{w,x\})\,\psi\big] \quad &\text{is} \quad g(f(x)) = x \wedge f(x) \neq w, \\
\mathrm{Sk}_{\{w\}}\big[\forall x(\exists y/\{w\})(\exists z/\{w,x\})\,\psi\big] \quad &\text{is} \quad \forall x\big[g(f(x)) = x \wedge f(x) \neq w\big], \\
\mathrm{Sk}(\varphi_\infty) \quad &\text{is} \quad \forall x\big[g(f(x)) = x \wedge f(x) \neq c\big],
\end{aligned}
$$

where f and g are fresh unary function symbols, and c is a fresh constant symbol. The Skolemization of φ_∞ asserts that f is an injection whose range is not the entire universe. Thus $\mathrm{Sk}(\varphi_\infty)$ is satisfiable by an expansion of $\mathbb{M}$ if and only if the universe of $\mathbb{M}$ is (Dedekind) infinite. ⊣

During the Skolemization process existential quantifiers are eliminated by introducing fresh function symbols. Sometimes we wish two of these function symbols to denote a single function. There is a standard trick we can use to ensure two function symbols have the same interpretation. Let φ be the IF sentence

$$\forall x \forall y \left(\exists u / \{y\}\right) \left(\exists v / \{x, u\}\right) \left[(x = y \rightarrow u = v) \wedge \psi\right].$$

The Skolem form of φ is

$$\forall x \forall y \left[\left(x = y \rightarrow f(x) = g(y)\right) \wedge \mathrm{Sk}_{\{x,y,u,v\}}(\psi)\right].$$

The conjunct $\left(x = y \rightarrow f(x) = g(y)\right)$ ensures that f and g have the same interpretation in any model of $\mathrm{Sk}(\varphi)$.

Example 4.15 An *involution* is a function f that satisfies $f\big(f(x)\big) = x$ for all x in its domain. A finite structure has an even number of elements if and only if there is a way of pairing the elements without leaving any element out, i.e., if there exists an involution without a fixed point. Let φ_{even} be the IF sentence

$$\forall x \forall y \left(\exists u / \{y\}\right) \left(\exists v / \{x, u\}\right)$$
$$\left[(x = y \rightarrow u = v) \wedge (u = y \rightarrow v = x) \wedge u \neq x\right].$$

The Skolem form of φ_{even} is

$$\forall x \forall y \left[\left(x = y \rightarrow f(x) = g(y)\right) \wedge \left(f(x) = y \rightarrow g(y) = x\right) \wedge f(x) \neq x\right].$$

Since f and g denote the same function, we can simplify $\mathrm{Sk}(\varphi_{\mathrm{even}})$ to

$$\forall x \left[f\big(f(x)\big) = x \wedge f(x) \neq x\right],$$

which asserts that f is an involution without a fixed point. Therefore $\mathrm{Sk}(\varphi_{\mathrm{even}})$ is satisfiable by an expansion of a finite structure if and only if the universe of the structure has an even number of elements.   ⊣

Example 4.16 A *graph* is a structure $\mathbb{G} = (V; E)$ where V is a set of *vertices* and E is a set of *edges*. An edge is normally thought of as an unordered pair $\{x, y\}$ for some $x, y \in V$, but in order to treat a graph as a structure in the sense of Definition 3.6, we take E to be a binary relation on V that is symmetric and irreflexive. A *matching* in $\mathbb{G}$ is a set M of edges such that no two edges in M share a vertex. A matching is *perfect* if every vertex is incident to an edge in M. A graph admits a perfect matching if and only if there is an involution f such that for every vertex x we have $E(x, f(x))$. Let φ_{PM} be the IF sentence

$$\forall x \forall y (\exists u / \{y\})(\exists v / \{x, u\})$$
$$\Big[(x = y \rightarrow u = v) \wedge (u = y \rightarrow v = x) \wedge E(x, u)\Big].$$

The Skolem form of φ_{PM} is

$$\forall x \forall y \Big[(x = y \rightarrow f(x) = g(y)) \wedge (f(x) = y \rightarrow g(y) = x) \wedge E(x, f(x))\Big],$$

which is equivalent to

$$\forall x \forall y \Big[f(f(x)) = x \wedge E(x, f(x))\Big].$$

Thus a graph satisfies φ_{PM} if and only if it admits a perfect matching. ⊣

Now let us a take a short excursion to the philosophy of language. In this field, games are used to model communication between agents. Lewis [38] defines a *signaling problem* as a situation in which an agent called the *communicator* wishes to communicate with one or more other agents called the *audience*. First, the communicator observes one of several alternative states of affairs, which the audience cannot directly observe. He then sends a signal to the audience. After receiving the signal, the audience performs one of several alternative actions, called *responses*. Every state of affairs x has a corresponding response $b(x)$ that the communicator and the audience agree is the best action to take when x holds. Lewis argues that a word acquires its meaning in virtue of its role in the solutions to various signaling problems.

Let S be a set of states of affairs, Σ a set of signals, and R a set of responses. Also let $b \colon S \rightarrow R$ be the function that sends each state of affairs to its best response.[2] The communicator employs an encoding $f \colon S \rightarrow \Sigma$ to choose a signal for every state of affairs. The audience employs a decoding $g \colon \Sigma \rightarrow R$ to decide which action to perform in response to the signal it receives. A *signaling system* is a pair (f, g) of encoding and decoding functions such that the composition $g \cdot f = b$ [38, pp. 130–132].

For example, imagine a driver who is trying to back into a parking space. Luckily, she has an assistant who gets out of the car and stands in a location where she can simultaneously see how much space there is behind the car and be seen by the driver. There are two states of affairs

---

[2] Lewis requires the function b to be a bijection, but we see no reason why two different states of affairs might not call for the same response.

the assistant wishes to communicate, i.e., whether there is enough space behind the car for the driver to continue backing up. The assistant has two signals at her disposal: she can stand palms facing in or palms facing out. The driver has two possible responses: she can back up or she can stop. Interestingly, there are two solutions to this signaling problem. The assistant can stand palms facing in when there is space, and palms facing out when there is no space, or vice versa. In the first case, the driver should continue backing up when she sees the assistant stand palms facing in, and stop when the assistant stands palms facing out. In the second case, the driver should stop when the assistant stands palms facing in, and back up when the assistant stands palms facing out. Both systems work equally well. If the assistant and the driver adopt mismatched encoding and decoding functions, however, an argument is bound to ensue.

The IF sentence $\forall x \exists z (\exists y / \{x\}) x = y$ from Example 4.11 can be modified to express a Lewis signaling game. In the following sentence φ, think of x as a situation, z as the signal sent by the sender, and y as the receiver's interpretation of the signal:

$$\forall x \exists z (\exists y / \{x\}) \Big[S(x) \rightarrow \big(\Sigma(z) \wedge R(y) \wedge y = b(x) \big) \Big].$$

The Skolem form of φ is

$$\forall x \Big[S(x) \rightarrow \big(\Sigma(f(x)) \wedge R(g(f(x))) \wedge g(f(x)) = b(x) \big) \Big].$$

If $\mathbb{M}$ is a suitable model for φ, then the signaling problem expressed by φ has a solution if and only if there is an expansion $\mathbb{M}^*$ of $\mathbb{M}$ such that $\mathbb{M}^* \models \text{Sk}(\varphi)$. Thus a signaling system is really just a pair of Skolem functions that encode a winning strategy for the semantic game of a certain IF sentence.

Conversely, one can think of the semantic game for any IF sentence as a generalized signaling problem in which each existential quantifier corresponds to an agent who can both send and receive signals, while the universal quantifiers correspond to states of affairs beyond the agents' control.

Skolem functions encode Eloise's strategies for the relevant semantic game. In the next section, we show how to use Kreisel's counterexamples to encode Abelard's strategies.

4.3.1 Falsity and Kreisel counterexamples

Definition 4.17   Let φ be an IF_L formula, and let

$$L^* = L \cup \left\{ f_\psi : \psi \in \mathrm{Subf}_\forall(\varphi) \right\}$$

be the expansion of L obtained by adding a fresh function symbol for every universally quantified subformula of φ. The *Kreisel form* (or *Kreiselization*) of $\psi \in \mathrm{Subf}(\varphi)$ with variables in U is defined recursively:

$$
\begin{aligned}
\mathrm{Kr}_U(\psi) &\;\; \text{is} \;\; \neg\psi \quad (\psi \text{ literal}), \\
\mathrm{Kr}_U(\psi \vee \psi') &\;\; \text{is} \;\; \mathrm{Kr}_U(\psi) \wedge \mathrm{Kr}_U(\psi'), \\
\mathrm{Kr}_U(\psi \wedge \psi') &\;\; \text{is} \;\; \mathrm{Kr}_U(\psi) \vee \mathrm{Kr}_U(\psi'), \\
\mathrm{Kr}_U\big((\exists x/W)\psi\big) &\;\; \text{is} \;\; \forall x\, \mathrm{Kr}_{U \cup \{x\}}(\psi), \\
\mathrm{Kr}_U\big((\forall x/W)\psi\big) &\;\; \text{is} \;\; \mathrm{Subst}\big(\mathrm{Kr}_{U \cup \{x\}}(\psi), x, f_{(\forall x/W)\psi}(y_1, \dots, y_n)\big),
\end{aligned}
$$

where $y_1, \dots, y_n$ enumerates the variables in $U - W$. An interpretation of $f_{(\forall x/W)\psi}$ is called a *Kreisel counterexample*. We abbreviate $\mathrm{Kr}_\varnothing(\varphi)$ by $\mathrm{Kr}(\varphi)$.   ⊣

The Kreisel form of φ is just the Skolem form of the dual of φ, which justifies the following definition.

Definition 4.18   Let φ be an IF_L formula, $\mathbb{M}$ a suitable structure, and s an assignment whose domain contains $\mathrm{Free}(\varphi)$. Define

$$\mathbb{M}, s \models^-_{\mathrm{Sk}} \varphi \quad \text{iff} \quad \mathbb{M}^*, s \models \mathrm{Kr}_{\mathrm{dom}(s)}(\varphi)$$

for some expansion $\mathbb{M}^*$ of $\mathbb{M}$ to the vocabulary

$$L^* = L \cup \left\{ f_\psi : \psi \in \mathrm{Subf}_\forall(\varphi) \right\}.$$ ⊣

Theorem 4.19   *Let φ be an IF_L formula, $\mathbb{M}$ a suitable structure, and s an assignment whose domain contains $\mathrm{Free}(\varphi)$. Then*

$$\mathbb{M}, s \models^-_{\mathrm{GTS}} \varphi \quad \text{iff} \quad \mathbb{M}, s \models^-_{\mathrm{Sk}} \varphi.$$

Proof   The proof is dual to the proof of Theorem 4.13.   ⊣

Example 4.20   Abelard does not have a winning strategy for the Matching Pennies sentence $\forall x (\exists y/\{x\}) x = y$ in any structure. The

Kreisel form helps us see why:

$$\mathrm{Kr}_{\{x,y\}}(x = y) \quad \text{is} \quad x \neq y,$$

$$\mathrm{Kr}_{\{x\}}\Big[(\exists y/\{x\})x = y\Big] \quad \text{is} \quad \forall y(x \neq y),$$

$$\mathrm{Kr}\Big[\forall x(\exists y/\{x\})x = y\Big] \quad \text{is} \quad \forall y(c \neq y),$$

where c is a fresh constant symbol. The Kreisel form of the Matching Pennies sentence is not true in any structure because the constant symbol c is interpreted by some element of the universe. ⊣

Example 4.21 In Example 4.8 we saw that Eloise has a winning strategy for the Matching Pennies sentence augmented with a dummy quantifier,

$$\forall x \exists y(\exists y/\{x\})x = y.$$

In Example 4.11 we saw that which variable we use for the dummy quantifier ($\exists y$ versus $\exists z$) does not affect the Skolem form. It does affect the Kreisel form:

$$\mathrm{Kr}_{\{x,y\}}(x = y) \quad \text{is} \quad x \neq y,$$

$$\mathrm{Kr}_{\{x,y\}}\Big[(\exists y/\{x\})x = y\Big] \quad \text{is} \quad \forall y(x \neq y),$$

$$\mathrm{Kr}_{\{x\}}\Big[\exists y(\exists y/\{x\})x = y\Big] \quad \text{is} \quad \forall y\forall y(x \neq y),$$

$$\mathrm{Kr}\Big[\forall x \exists y(\exists y/\{x\})x = y\Big] \quad \text{is} \quad \forall y\forall y(c \neq y),$$

versus

$$\mathrm{Kr}_{\{x,y,z\}}(x = y) \quad \text{is} \quad x \neq y,$$

$$\mathrm{Kr}_{\{x,z\}}\Big[(\exists y/\{x\})x = y\Big] \quad \text{is} \quad \forall y(x \neq y),$$

$$\mathrm{Kr}_{\{x\}}\Big[\exists z(\exists y/\{x\})x = y\Big] \quad \text{is} \quad \forall z\forall y(x \neq y),$$

$$\mathrm{Kr}\Big[\forall x \exists z(\exists y/\{x\})x = y\Big] \quad \text{is} \quad \forall z\forall y(c \neq y).$$

However, observe that both Kreisel forms are equivalent to $\forall y(c \neq y)$. ⊣

4.4 Compositional semantics

A semantics is said to be compositional if the meaning of a formula is determined by the meanings of its subformulas and the manner of their composition. Such semantics are pleasant to work with because they enable us to calculate the meanings of formulas in a systematic way, starting from the meanings of atomic formulas (or literals, in the case of IF logic). So far we have defined two semantics for IF logic, neither of which is compositional because they do not define the meaning of an IF formula in terms of the meanings of its parts. In this section we present a compositional semantics for IF logic called *trump semantics* discovered by Hodges [32, 33].

Compositionality does not come for free. In order to encode enough information about IF formulas to make the semantics compositional, we must switch from using single assignments to sets of assignments.

Definition 4.22 A *team of assignments* (or *assignment team*) in $\mathbb{M}$ is a set of assignments in $\mathbb{M}$ that have a common domain. If X is a nonempty assignment team then the *domain* of X, written $\mathrm{dom}(X)$, is the common domain of the assignments in X. For technical reasons, we take the domain of the empty team of assignments to be set of all variables. ⊣

Definition 4.23 Let X be a team of assignments in a structure $\mathbb{M}$. Also let $a \in M$, $A \subseteq M$, and $f \colon X \to A$. We define

$$X[x, a] = \big\{ s(x/a) : s \in X \big\},$$
$$X[x, A] = \big\{ s(x/a) : s \in X, \, a \in A \big\},$$
$$X[x, f] = \big\{ s(x/f(s)) : s \in X \big\}. \qquad \dashv$$

Given two assignments s and s' we say that s' *extends* s if $s \subseteq s'$. Given two assignment teams X and Y, we say that Y *extends* X if every $s \in X$ has an extension $s' \in Y$, and every $s' \in Y$ is an extension of some $s \in X$. If $x \notin \mathrm{dom}(X)$, then $X[x, f]$ is a minimal extension of X to $\mathrm{dom}(X) \cup \{x\}$, while $X[x, M]$ is the maximal extension of X to $\mathrm{dom}(X) \cup \{x\}$. A *Cartesian extension* of X has the form

$$X \times Y = \{s \cup s' : s \in X, \, s' \in Y\},$$

where $\mathrm{dom}(X) \cap \mathrm{dom}(Y) = \varnothing$.

Recall that we write $s \approx_W s'$ when s and s' are two assignments with the same domain that agree on every variable in $\mathrm{dom}(s) - W$.

Definition 4.24 Let X be a team of assignments in $\mathbb{M}$ such that $W \subseteq \mathrm{dom}(X)$. A function $f \colon X \to M$ is *uniform in W* (or W-*uniform*) if for all $s, s' \in X$,

$$s \approx_W s' \quad \text{implies} \quad f(s) = f(s'). \qquad \dashv$$

A *cover* of a set X is a family of sets whose union is X. When $\{Y, Y'\}$ is a cover of X, we simply write $Y \cup Y' = X$.

We are now ready to define a compositional semantics for IF logic in terms of teams of assignments. We will see later (Section 8.1) that one cannot define a compositional semantics for IF logic in terms of single assignments.

Definition 4.25 Let φ and φ' be IF formulas, $\mathbb{M}$ a suitable structure, and X a team of assignments in $\mathbb{M}$ whose domain contains Free(φ) and Free(φ'). If φ is a literal, define

$$\mathbb{M}, X \models^+_{\text{Tr}} \varphi \quad \text{iff} \quad \mathbb{M}, s \models \varphi \qquad \text{(for all } s \in X\text{)}.$$

Furthermore,

$$\mathbb{M}, X \models^+_{\text{Tr}} \varphi \vee \varphi' \quad \text{iff} \quad \mathbb{M}, Y \models^+_{\text{Tr}} \varphi \text{ and } \mathbb{M}, Y' \models^+_{\text{Tr}} \varphi'$$
$$\text{for some } Y \cup Y' = X;$$

$$\mathbb{M}, X \models^+_{\text{Tr}} \varphi \wedge \varphi' \quad \text{iff} \quad \mathbb{M}, X \models^+_{\text{Tr}} \varphi \text{ and } \mathbb{M}, X \models^+_{\text{Tr}} \varphi';$$

$$\mathbb{M}, X \models^+_{\text{Tr}} (\exists x/W)\varphi \quad \text{iff} \quad \mathbb{M}, X[x, f] \models^+_{\text{Tr}} \varphi$$
$$\text{for some } W\text{-uniform function } f \colon X \to M;$$

$$\mathbb{M}, X \models^+_{\text{Tr}} (\forall x/W)\varphi \quad \text{iff} \quad \mathbb{M}, X[x, M] \models^+_{\text{Tr}} \varphi.$$

When $\mathbb{M}, X \models^+_{\text{Tr}} \varphi$ we say that $\mathbb{M}, X$ *satisfies* φ, and that X is a *winning team of assignments for* φ *in* $\mathbb{M}$.[3]

We write $\mathbb{M}, s \models^+_{\text{Tr}} \varphi$ as an abbreviation for $\mathbb{M}, \{s\} \models^+_{\text{Tr}} \varphi$, unless s is the empty assignment. To avoid confusion, we abbreviate $\mathbb{M}, \{\varnothing\} \models^+_{\text{Tr}} \varphi$ by $\mathbb{M} \models^+_{\text{Tr}} \varphi$, and write $\mathbb{M}, \varnothing \models^+_{\text{Tr}} \varphi$ to indicate that the empty team of assignments satisfies φ.[4] $\dashv$

When $x \notin \text{dom}(X)$, we can say that X satisfies $(\exists x/W)\varphi$ if and only if some minimal extension of X to $\text{dom}(X) \cup \{x\}$ satisfies φ. Dually, X satisfies $(\forall x/W)\varphi$ if and only if the maximal extension of X satisfies φ.

In Example 4.8 we saw that Eloise can use the extra quantifier $\exists y$ in

$$\forall x \exists y (\exists y/\{x\}) x = y$$

---

[3] Hodges calls a nonempty winning team of assignments a *trump*; hence the name *trump semantics* [32, p. 552].

[4] We shall see that $\mathbb{M}, \varnothing \models^+_{\text{Tr}} \varphi$ always holds.

to signal the value of x to herself. Let us examine how signaling manifests itself in the trump semantics of this sentence.

Example 4.26 Let $\mathbb{M} = \{a, b\}$ be a structure with two elements. Let $s_a = \{(x, a)\}$ and $s_b = \{(x, b)\}$. To prove $\mathbb{M} \models^+_{\mathrm{Tr}} \forall x \exists y (\exists y / \{x\}) x = y$, it suffices to show that

$$\mathbb{M}, \{s_a, s_b\} \models^+_{\mathrm{Tr}} \exists y (\exists y / \{x\}) x = y.$$

Let $f(s_a) = a$ and $f(s_b) = b$. Then $\{s_a, s_b\}[y, f] = \{s_{aa}, s_{bb}\}$, where

$$s_{aa} = \{(x, a), (y, a)\} \quad \text{and} \quad s_{bb} = \{(x, b), (y, b)\}.$$

To show $\mathbb{M}, \{s_{aa}, s_{bb}\} \models^+_{\mathrm{Tr}} (\exists y / \{x\}) x = y$, let $g(s_{aa}) = a$ and $g(s_{bb}) = b$. Observe that g is $\{x\}$-uniform, $\{s_{aa}, s_{bb}\}[y, g] = \{s_{aa}, s_{bb}\}$, and

$$\mathbb{M}, \{s_{aa}, s_{bb}\} \models^+_{\mathrm{Tr}} x = y.$$

Now let us see how changing the dummy quantifier from $\exists y$ to $\exists z$ affects the trump semantics. Let

$$s'_{aa} = \{(x, a), (z, a)\}, \qquad s_{aaa} = \{(x, a), (y, a), (z, a)\},$$
$$s'_{bb} = \{(x, b), (z, b)\}, \qquad s_{bbb} = \{(x, b), (y, b), (z, b)\}.$$

Define $g'(s'_{aa}) = a$ and $g'(s'_{bb}) = b$. Then

$$\{s_a, s_b\}[z, f] = \{s'_{aa}, s'_{bb}\},$$
$$\{s'_{aa}, s'_{bb}\}[y, g'] = \{s_{aaa}, s_{bbb}\},$$

and $\mathbb{M}, \{s_{aaa}, s_{bbb}\} \models^+_{\mathrm{Tr}} x = y$. Since g' is uniform in $\{x\}$, it follows that

$$\mathbb{M}, \{s'_{aa}, s'_{bb}\} \models^+_{\mathrm{Tr}} (\exists y / \{x\}) x = y,$$
$$\mathbb{M}, \{s_a, s_b\} \models^+_{\mathrm{Tr}} \exists z (\exists y / \{x\}) x = y,$$
$$\mathbb{M} \models^+_{\mathrm{Tr}} \forall x \exists z (\exists y / \{x\}) x = y. \qquad \dashv$$

The previous example highlights an important difference between first-order logic and IF logic. Adding a vacuous quantifier (or quantifying the same variable twice) does not affect the meaning of a first-order formula, but it can affect the meaning of an IF formula. For example, if φ is the IF formula $(\exists y / \{x\}) x = y$ then $\mathbb{M}, \{s_a, s_b\} \not\models^+_{\mathrm{Tr}} \varphi$, while

$$\mathbb{M}, \{s_a, s_b\} \models^+_{\mathrm{Tr}} \exists y \varphi \quad \text{and} \quad \mathbb{M}, \{s_a, s_b\} \models^+_{\mathrm{Tr}} \exists z \varphi.$$

To prove that trump semantics is equivalent to Skolem semantics, we must extend Skolem semantics to teams of assignments.

Definition 4.27 Let φ be an IF formula, $\mathbb{M}$ a suitable structure, and X a team of assignments whose domain contains $\mathrm{Free}(\varphi)$. We define $\mathbb{M}, X \models^+_{\mathrm{Sk}} \varphi$ to mean there is an expansion $\mathbb{M}^*$ of $\mathbb{M}$ to the vocabulary

$$L^* = L \cup \left\{ f_\psi : \psi \in \mathrm{Subf}_\exists(\varphi) \right\}$$

such that for all $s \in X$ we have $\mathbb{M}^*, s \models \mathrm{Sk}_{\mathrm{dom}(X)}(\psi)$.   ⊣

Theorem 4.28 *Let φ be an IF_L formula, $\mathbb{M}$ a suitable structure, and X a team of assignments whose domain contains $\mathrm{Free}(\varphi)$. Then*

$$\mathbb{M}, X \models^+_{\mathrm{Tr}} \varphi \quad \textit{iff} \quad \mathbb{M}, X \models^+_{\mathrm{Sk}} \varphi.$$

Proof We prove by induction that for every $\psi \in \mathrm{Subf}(\varphi)$ and every assignment team Y whose domain contains both $\mathrm{dom}(X)$ and $\mathrm{Free}(\psi)$,

$$\mathbb{M}, Y \models^+_{\mathrm{Tr}} \psi \quad \text{iff} \quad \mathbb{M}, Y \models^+_{\mathrm{Sk}} \psi.$$

The basis step follows easily from the definitions.

Suppose ψ is $\psi_1 \vee \psi_2$. If $\mathbb{M}, Y \models^+_{\mathrm{Tr}} \psi_1 \vee \psi_2$ then $\mathbb{M}, Y_1 \models^+_{\mathrm{Tr}} \psi_1$ and $\mathbb{M}, Y_2 \models^+_{\mathrm{Tr}} \psi_2$ for some cover $Y_1 \cup Y_2 = Y$. By inductive hypothesis, there is an expansion $\mathbb{M}_1$ of $\mathbb{M}$ to the vocabulary

$$L_1 = L \cup \left\{ f_\chi : \chi \in \mathrm{Subf}_\exists(\psi_1) \right\}$$

such that for all $s \in Y_1$ we have $\mathbb{M}_1, s \models \mathrm{Sk}_{\mathrm{dom}(Y_1)}(\psi_1)$. Likewise, there is an expansion $\mathbb{M}_2$ of $\mathbb{M}$ to the vocabulary

$$L_2 = L \cup \left\{ f_\chi : \chi \in \mathrm{Subf}_\exists(\psi_2) \right\}$$

such that for all $s \in Y_2$ we have $\mathbb{M}_2, s \models \mathrm{Sk}_{\mathrm{dom}(Y_2)}(\psi_2)$. We may assume $L_1 \cap L_2 = L$, so there is a common expansion $\mathbb{M}^*$ to the vocabulary

$$L^* = L \cup \left\{ f_\chi : \chi \in \mathrm{Subf}_\exists(\psi) \right\}$$

such that for all $s \in Y$,

$$\mathbb{M}^*, s \models \mathrm{Sk}_{\mathrm{dom}(Y_1)}(\psi_1) \quad \text{or} \quad \mathbb{M}^*, s \models \mathrm{Sk}_{\mathrm{dom}(Y_2)}(\psi_2),$$

which implies $\mathbb{M}^*, s \models \mathrm{Sk}_{\mathrm{dom}(Y)}(\psi_1 \vee \psi_2)$. Therefore $\mathbb{M}, Y \models^+_{\mathrm{Sk}} \psi_1 \vee \psi_2$.

Conversely, suppose there is an expansion $\mathbb{M}^*$ of $\mathbb{M}$ to the vocabulary L^* such that for all $s \in Y$ we have $\mathbb{M}^*, s \models \mathrm{Sk}_{\mathrm{dom}(Y)}(\psi_1 \vee \psi_2)$. Let

$$Y_i = \left\{ s \in Y : \mathbb{M}^*, s \models \mathrm{Sk}_{\mathrm{dom}(Y_i)}(\psi_i) \right\}.$$

Then $Y_1 \cup Y_2 = Y$. In addition, we have $\mathbb{M}, Y_1 \models^+_{\mathrm{Sk}} \psi_1$ and $\mathbb{M}, Y_2 \models^+_{\mathrm{Sk}} \psi_2$, so by inductive hypothesis $\mathbb{M}, Y_1 \models^+_{\mathrm{Tr}} \psi_1$ and $\mathbb{M}, Y_2 \models^+_{\mathrm{Tr}} \psi_2$. Thus $\mathbb{M}, Y \models^+_{\mathrm{Tr}} \psi_1 \vee \psi_2$.

Suppose ψ is $\psi_1 \wedge \psi_2$. If $\mathbb{M}, Y \models^+_{\mathrm{Tr}} \psi_1 \wedge \psi_2$, then $\mathbb{M}, Y \models^+_{\mathrm{Tr}} \psi_1$ and $\mathbb{M}, Y \models^+_{\mathrm{Tr}} \psi_2$, so by inductive hypothesis there are expansions $\mathbb{M}_1$ and $\mathbb{M}_2$ to the vocabularies L_1 and L_2, respectively, such that for all $s \in Y$ we have $\mathbb{M}_1, s \models \mathrm{Sk}_{\mathrm{dom}(Y)}(\psi_1)$ and $\mathbb{M}_2, s \models \mathrm{Sk}_{\mathrm{dom}(Y)}(\psi_2)$. Hence there is a common expansion $\mathbb{M}^*$ to the vocabulary L^* such that for all $s \in Y$,

$$\mathbb{M}^*, s \models \mathrm{Sk}_{\mathrm{dom}(Y)}(\psi_1) \quad \text{and} \quad \mathbb{M}^*, s \models \mathrm{Sk}_{\mathrm{dom}(Y)}(\psi_2),$$

which implies

$$\mathbb{M}^*, s \models \mathrm{Sk}_{\mathrm{dom}(Y)}(\psi_1 \wedge \psi_2).$$

Thus $\mathbb{M}, Y \models^+_{\mathrm{Sk}} \psi_1 \wedge \psi_2$.

Conversely, suppose there exists an expansion $\mathbb{M}^*$ of $\mathbb{M}$ to the vocabulary L^* such that for all $s \in Y$ we have $\mathbb{M}^*, s \models \mathrm{Sk}_{\mathrm{dom}(Y)}(\psi_1 \wedge \psi_2)$. Let $\mathbb{M}_i$ be the reduct of $\mathbb{M}^*$ to the vocabulary L_i. Then for all $s \in Y$,

$$\mathbb{M}_1, s \models \mathrm{Sk}_{\mathrm{dom}(Y)}(\psi_1) \quad \text{and} \quad \mathbb{M}_2, s \models \mathrm{Sk}_{\mathrm{dom}(Y)}(\psi_2),$$

which implies $\mathbb{M}, Y \models^+_{\mathrm{Sk}} \psi_1$ and $\mathbb{M}, Y \models^+_{\mathrm{Sk}} \psi_2$. By inductive hypothesis

$$\mathbb{M}, Y \models^+_{\mathrm{Tr}} \psi_1 \quad \text{and} \quad \mathbb{M}, Y \models^+_{\mathrm{Tr}} \psi_2,$$

which implies $\mathbb{M}, Y \models^+_{\mathrm{Tr}} \psi_1 \wedge \psi_2$.

Suppose ψ is $(\exists x/W)\psi'$. If $\mathbb{M}, Y \models^+_{\mathrm{Tr}} (\exists x/W)\psi'$ there is a W-uniform $f : Y \to M$ such that $\mathbb{M}, Y[x, f] \models^+_{\mathrm{Tr}} \psi'$. By inductive hypothesis

$$\mathbb{M}, Y[x, f] \models^+_{\mathrm{Sk}} \psi',$$

so there is an expansion $\mathbb{M}'$ of $\mathbb{M}$ to the vocabulary

$$L' = L \cup \{\, f_\chi : \chi \in \mathrm{Subf}_\exists(\psi') \,\}$$

such that for all $s \in Y$ we have

$$\mathbb{M}', s\big(x/f(s)\big) \models \mathrm{Sk}_{\mathrm{dom}(Y) \cup \{x\}}(\psi').$$

Let $\mathbb{M}^*$ be an expansion of $\mathbb{M}'$ to the vocabulary L^* such that for all $s \in Y$,

$$f^{\mathbb{M}^*}_{(\exists x/W)\psi'}\big(s(y_1), \ldots, s(y_n)\big) = f(s),$$

where $y_1, \ldots, y_n$ enumerates $\mathrm{dom}(Y) - W$. Observe that $\mathbb{M}^*$ is well defined because f is uniform in W. Then for all $s \in Y$,

$$\mathbb{M}^*, s \models \mathrm{Subst}\big(\mathrm{Sk}_{\mathrm{dom}(Y) \cup \{x\}}(\psi'), x, f_{(\exists x/W)\psi'}(y_1, \ldots, y_n)\big)$$

by the substitution lemma. Hence $\mathbb{M}^*, s \models \mathrm{Sk}_{\mathrm{dom}(Y)}\big((\exists x/W)\psi'\big)$. Thus $\mathbb{M}, Y \models^+_{\mathrm{Sk}} (\exists x/W)\psi'$.

Conversely, suppose there is an expansion $\mathbb{M}^*$ of $\mathbb{M}$ to the vocabulary L^* such that for all $s \in Y$,

$$\mathbb{M}^*, s \models \text{Subst}\big(\text{Sk}_{\text{dom}(Y) \cup \{x\}}(\psi'), x, f_{(\exists x/W)\psi'}(y_1, \ldots, y_n)\big).$$

Define a W-uniform function $f \colon Y \to M$ by

$$f(s) = f^{\mathbb{M}^*}_{(\exists x/W)\psi'}\big(s(y_1), \ldots, s(y_n)\big),$$

where $y_1, \ldots, y_n$ enumerates $\text{dom}(Y) - W$, and let $\mathbb{M}'$ be the reduct of $\mathbb{M}^*$ to the vocabulary L'. Then for all $s \in Y$,

$$\mathbb{M}', s\big(x/f(s)\big) \models \text{Sk}_{\text{dom}(Y) \cup \{x\}}(\psi'),$$

which implies $\mathbb{M}, Y[x, f] \models^+_{\text{Sk}} \psi'$. By inductive hypothesis

$$\mathbb{M}, Y[x, f] \models^+_{\text{Tr}} \psi'.$$

Thus $\mathbb{M}, X \models^+_{\text{Tr}} (\exists x/W)\psi'$.

Suppose ψ is $(\forall x/W)\psi'$. If $\mathbb{M}, Y \models^+_{\text{Tr}} (\forall x/W)\psi'$, then

$$\mathbb{M}, Y[x, M] \models^+_{\text{Tr}} \psi'.$$

By inductive hypothesis $\mathbb{M}, Y[x, M] \models^+_{\text{Sk}} \psi'$, so there is an expansion $\mathbb{M}^*$ of $\mathbb{M}$ to the vocabulary L^* such that for all $s \in Y$ and $a \in M$,

$$\mathbb{M}^*, s(x/a) \models \text{Sk}_{\text{dom}(Y) \cup \{x\}}(\psi').$$

It follows that for all $s \in Y$ we have $\mathbb{M}^*, s \models \forall x\, \text{Sk}_{\text{dom}(Y) \cup \{x\}}(\psi')$. Thus $\mathbb{M}, Y \models^+_{\text{Sk}} (\forall x/W)\psi'$.

Conversely, suppose there is an expansion $\mathbb{M}^*$ of $\mathbb{M}$ to the vocabulary L^* such that for all $s \in Y$ we have $\mathbb{M}^*, s \models \forall x\, \text{Sk}_{\text{dom}(Y) \cup \{x\}}\psi'$. Then for all $s \in Y$ and $a \in M$,

$$\mathbb{M}^*, s(x/a) \models \text{Sk}_{\text{dom}(Y) \cup \{x\}}(\psi').$$

Hence $\mathbb{M}, Y[x, M] \models^+_{\text{Sk}} \psi'$. By inductive hypothesis $\mathbb{M}, Y[x, M] \models^+_{\text{Tr}} \psi'$, which implies $\mathbb{M}, Y \models^+_{\text{Tr}} (\forall x/W)\psi'$. ⊣

So far we have only defined the "truth" half of trump semantics. The "falsity" half is its mirror image. The falsity clause for literals is

$$\mathbb{M}, X \models^-_{\text{Tr}} \varphi \quad \text{iff} \quad \mathbb{M}, s \not\models \varphi, \text{ for all } s \in X.$$

The other falsity clauses are obtained from the truth clauses in Definition 4.25 by exchanging everywhere $\vee$ with $\wedge$ and $(\exists x/W)$ with $(\forall x/W)$. When $\mathbb{M}, X \models^-_{\text{Tr}} \varphi$ we say that $\mathbb{M}, X$ *dissatisfies* φ, and that X is a *losing team of assignments for φ in $\mathbb{M}$*.

We adopt the same convention concerning singleton teams as before. If s is not the empty assignment,

$$\mathbb{M}, s \models_{\mathrm{Tr}}^{-} \varphi \quad \text{iff} \quad \mathbb{M}, \{s\} \models_{\mathrm{Tr}}^{-} \varphi,$$

and for IF sentences

$$\mathbb{M} \models_{\mathrm{Tr}}^{-} \varphi \quad \text{iff} \quad \mathbb{M}, \{\varnothing\} \models_{\mathrm{Tr}}^{-} \varphi.$$

We can extend the Skolem semantics for falsity to assignment teams by declaring $\mathbb{M}, X \models_{\mathrm{Sk}}^{-} \varphi$ if and only if there is an expansion $\mathbb{M}^*$ to the vocabulary

$$L^* = L \cup \{ f_\psi : \psi \in \mathrm{Subf}_\forall(\psi) \}$$

such that for all $s \in X$ we have $\mathbb{M}^*, s \models \mathrm{Kr}_{\mathrm{dom}(X)}(\varphi)$. We leave the proof of the next theorem as an exercise for the reader.

Theorem 4.29 *Let φ be an IF formula, $\mathbb{M}$ a suitable structure, and X a team of assignments whose domain contains* Free(φ). *Then*

$$\mathbb{M}, X \models_{\mathrm{Tr}}^{-} \varphi \quad \text{iff} \quad \mathbb{M}, X \models_{\mathrm{Sk}}^{-} \varphi. \hspace{2cm} \dashv$$

4.5 Game-theoretic semantics redux

We have defined the syntax for independence-friendly logic and three semantic interpretations. The first two interpretations were defined relative to single assignments, whereas the third interpretation was defined relative to teams of assignments. By identifying assignments with singleton teams, we proved that all three interpretations agree:

$$\mathbb{M}, s \models_{\mathrm{GTS}}^{+} \varphi \quad \text{iff} \quad \mathbb{M}, s \models_{\mathrm{Sk}}^{+} \varphi \quad \text{iff} \quad \mathbb{M}, s \models_{\mathrm{Tr}}^{+} \varphi,$$
$$\mathbb{M}, s \models_{\mathrm{GTS}}^{-} \varphi \quad \text{iff} \quad \mathbb{M}, s \models_{\mathrm{Sk}}^{-} \varphi \quad \text{iff} \quad \mathbb{M}, s \models_{\mathrm{Tr}}^{-} \varphi.$$

In order to prove the equivalence, we needed to extend Skolem semantics to teams of assignments. We can also extend game-theoretic semantics to assignment teams by allowing the initial assignment to vary from one play to another.

Definition 4.30 *Let φ be an IF formula, $\mathbb{M}$ a suitable structure, and X a team of assignments whose domain contains* Free(φ). *The semantic game $G(\mathbb{M}, X, \varphi)$ is defined exactly like $G(\mathbb{M}, s, \varphi)$ except that the initial assignment of a play may be any $s \in X$. To wit,*

- $H = \bigcup \{ H_\psi : \psi \in \mathrm{Subf}(\varphi) \}$, where H_ψ is defined recursively:
  - $H_\varphi = \{ (s, \varphi) : s \in X \}$,
  - if ψ is $\chi_1 \circ \chi_2$, then $H_{\chi_i} = \{ h^\frown \chi_i : h \in H_{\chi_1 \circ \chi_2} \}$,
  - if ψ is $(Qx/W)\chi$, then $H_\chi = \{ h^\frown(x, a) : h \in H_{(Qx/W)\chi}, a \in M \}$.

  Every history h' induces an assignment $s_{h'}$ extending and/or modifying the initial assignment:

$$s_{h'} = \begin{cases} s & \text{if } h' = (s, \varphi), \\ s_h & \text{if } h' = h^\frown \chi, \\ s_h(x/a) & \text{if } h' = h^\frown(x, a). \end{cases}$$

- Once play reaches a literal, the game ends:

$$Z = \bigcup \{ H_\chi : \chi \in \mathrm{Lit}(\varphi) \}.$$

- Disjunctions and existential quantifiers are decision points for Eloise, while conjunctions and universal quantifiers are decision points for Abelard:

$$P(h) = \begin{cases} \exists & \text{if } h \in H_{\chi \vee \chi'} \text{ or } h \in H_{(\exists x/W)\chi}, \\ \forall & \text{if } h \in H_{\chi \wedge \chi'} \text{ or } h \in H_{(\forall x/W)\chi}. \end{cases}$$

- For $h, h' \in H_{(\exists x/W)\chi}$ we have $h \sim_\exists h'$ if and only if $s_h \approx_W s_{h'}$.
  For $h, h' \in H_{(\forall x/W)\chi}$ we have $h \sim_\forall h'$ if and only if $s_h \approx_W s_{h'}$.

- Eloise wins if the literal χ reached at the end of play is satisfied by the current assignment; Abelard wins if it is not:

$$u(h) = \begin{cases} \exists & \text{if } \mathbb{M}, s_h \models \chi, \\ \forall & \text{if } \mathbb{M}, s_h \not\models \chi. \end{cases} \qquad \dashv$$

Notice that the histories of $G(\mathbb{M}, X, \varphi)$ form a forest rather than a tree (unless X is a singleton) because every $(s, \varphi) \in H_\varphi$ is a root. Let $H^{(s)}$ denote the set of histories for $G(\mathbb{M}, X, \varphi)$ in which the initial assignment is s. Note that if $s, s' \in X$ are distinct, then $H^{(s)}$ is disjoint from $H^{(s')}$ and

$$H = \bigcup \{ H^{(s)} : s \in X \}.$$

Playing an extensive game whose histories form a forest is similar to playing a game whose histories form a tree. The only difference is that the players may be unsure of the initial position.[5] For example, the goal

---

[5] Hodges calls a semantic game in which the initial assignment can vary a *contest* [32, p. 549].

of the popular game Minesweeper is to clear an abstract minefield by flagging all the mines. At first, the player has no idea where the mines are located. After a few moves, however, the player can start to deduce the location of the mines from the feedback provided by the game. At the beginning of the semantic game $G(\mathbb{M}, X, \varphi)$ the players are unsure which values the free variables of φ take. They know that the initial assignment belongs to X, but they are unaware of precisely which member of X is the actual assignment. It may help the reader's intuition to imagine that the initial assignment is secretly chosen from X by a disinterested third party (Nature).

Our definition of a strategy for an extensive game did not depend on there being a unique initial history (see Definition 2.4). Eloise or Abelard might have a winning strategy for $G(\mathbb{M}, X, \varphi)$, despite the lack of perfect information at the beginning of the game.

Definition 4.31 Let φ be an IF formula, $\mathbb{M}$ a suitable structure, and X a team of assignments whose domain contains $\text{Free}(\varphi)$.

$\mathbb{M}, X \models^+_{\text{GTS}} \varphi$ iff Eloise has a winning strategy for $G(\mathbb{M}, X, \varphi)$.

$\mathbb{M}, X \models^-_{\text{GTS}} \varphi$ iff Abelard has a winning strategy for $G(\mathbb{M}, X, \varphi)$. ⊣

The proof of Theorem 4.13 can be easily modified to show that

$$\mathbb{M}, X \models^+_{\text{GTS}} \varphi \quad \text{iff} \quad \mathbb{M}, X \models^+_{\text{Sk}} \varphi,$$

and the dual follows easily as well. Thus

$$\mathbb{M}, X \models^\pm_{\text{GTS}} \varphi \quad \text{iff} \quad \mathbb{M}, X \models^\pm_{\text{Sk}} \varphi \quad \text{iff} \quad \mathbb{M}, X \models^\pm_{\text{Tr}} \varphi.$$

Throughout the remainder of the book we shall use all three semantics and the terminology that comes with them interchangeably. We will use the symbols $\models^+$ and $\models^-$ instead of specifying a particular semantics $\models^\pm_{\text{GTS}}$, $\models^\pm_{\text{Sk}}$ or $\models^\pm_{\text{Tr}}$.

To conclude the chapter, let us consider another kind of signaling that can occur with IF formulas. We know that the players can use quantifiers to signal the values of hidden variables to themselves. They can also use connectives to accomplish the same task. In the next example, Eloise will use a disjunction to separate assignments based on the value they assign to x.

Example 4.32 Let φ be the formula

$$(\exists y/\{x\})x = y \vee (\exists y/\{x\})x = y.$$

Let $M = \{a, b\}$ and the various assignments be as in Example 4.26. We show that $M, \{s_a, s_b\} \models^+ \varphi$, but $M, \{s_a, s_b\} \not\models^+ (\exists y/\{x\})x = y$.

Suppose for the sake of a contradiction that

$$M, \{s_a, s_b\} \models^+ (\exists y/\{x\})x = y.$$

Then there is an $\{x\}$-uniform function $f \colon \{s_a, s_b\} \to M$ such that

$$M, \left\{ s_a\big(y/f(s_a)\big),\, s_b\big(y/f(s_b)\big) \right\} \models^+ x = y.$$

Since f is uniform in $\{x\}$ we must have

$$a = f(s_a) = f(s_b) = b,$$

a contradiction.

Now observe that $M, \{s_{aa}\} \models^+ x = y$ and $M, \{s_{bb}\} \models^+ x = y$. The functions $f \colon \{s_a\} \to M$ and $g \colon \{s_b\} \to M$ defined by $f(s_a) = a$ and $g(s_b) = b$ are both uniform in $\{x\}$. Moreover, $\{s_a\}[y, f] = \{s_{aa}\}$ and $\{s_b\}[y, g] = \{s_{bb}\}$. Thus

$$M, \{s_a\} \models^+ (\exists y/\{x\})x = y \quad \text{and} \quad M, \{s_b\} \models^+ (\exists y/\{x\})x = y.$$

Therefore $M, \{s_a, s_b\} \models^+ \varphi$. $\quad\dashv$

Because she can see the value of x when choosing the left or right disjunct, Eloise can choose the left disjunct whenever $s(x) = a$, and the right disjunct whenever $s(x) = b$. Thereafter, Eloise can infer the value of x from the current subformula.

5

Properties of IF logic

Many basic properties of IF logic were observed by Hodges while he was developing trump semantics [32, 33]. Subsequently, Caicedo and Krynicki attempted to prove a prenex normal form theorem for IF logic [10], but they failed to properly account for the subtleties of signaling. Later Caicedo, Dechesne, and Janssen succeeded in proving many logical equivalences for IF logic, including a prenex normal form theorem [9]. Additional equivalences and entailments first appeared in [39] and were later published in [40]. Basic model-theoretic properties of IF logic have been investigated in [24, 49–51].

5.1 Basic properties

A team of assignments encodes a player's knowledge about the current assignment. When we write $\mathbb{M}, X \models^+ \varphi$ we are asserting that Eloise has a winning strategy for the semantic game for φ as long as she knows the initial assignment belongs to X. Now imagine that, at the beginning of the game, an oracle informs Eloise that the initial assignment belongs to a subteam $Y \subset X$. Then Eloise gains an advantage because she has fewer possibilities to consider. Thus smaller teams represent more information about the current assignment. At the extremes, a singleton team represents having perfect information about the current assignment, while the team of all possible assignments with a given domain represents having no information about the current assignment.

The following propositions record two important properties of assignment teams: (1) every subteam of a winning team of assignments is winning, and every subteam of a losing team of assignments is losing; (2) the empty team of assignments is both winning and losing for every

IF formula, and it is the only team of assignments that can be both winning and losing for the same formula.[1]

Proposition 5.1 (Downward monotonicity) *Let φ be an IF formula, $\mathbb{M}$ a suitable structure, and X a team of assignments whose domain contains* Free(φ). *For every assignment team $Y \subseteq X$,*

$$\mathbb{M}, X \models^{\pm} \varphi \quad \text{implies} \quad \mathbb{M}, Y \models^{\pm} \varphi.$$

Proof Suppose Eloise has a winning strategy σ for $G(\mathbb{M}, X, \varphi)$ and $Y \subseteq X$. Let $H_\exists$ denote the histories of $G(\mathbb{M}, X, \varphi)$ in which it is Eloise's turn to move, and let $H'_\exists$ denote the histories of $G(\mathbb{M}, Y, \varphi)$ in which it is her move. We can obtain a strategy σ' for $G(\mathbb{M}, Y, \varphi)$ simply by restricting σ to $H'_\exists$. Every terminal history of $G(\mathbb{M}, Y, \varphi)$ in which Eloise follows σ' is a terminal history of $G(\mathbb{M}, X, \varphi)$ in which she follows σ. Therefore σ' is a winning strategy for $G(\mathbb{M}, Y, \varphi)$.

Similarly, if Abelard has a winning strategy τ for $G(\mathbb{M}, X, \varphi)$ then he has a winning strategy τ' for $G(\mathbb{M}, Y, \varphi)$. ⊣

Proposition 5.2 (Noncontradiction) *Let φ be an IF formula, $\mathbb{M}$ a suitable structure, and X a team of assignments whose domain contains* Free(φ). *Then $\mathbb{M}, X \models^+ \varphi$ and $\mathbb{M}, X \models^- \varphi$ if and only if $X = \varnothing$.*

Proof The semantic game $G(\mathbb{M}, \varnothing, \varphi)$ has no histories, so the empty strategy is winning for both players. If X is nonempty it is impossible for both players to have a winning strategy for $G(\mathbb{M}, X, \varphi)$. ⊣

That the empty team of assignments is both winning and losing for every IF formula may seem anomalous, but it is necessary to properly interpret disjunctions and conjunctions. For any two IF formulas, φ and ψ,

$$\mathbb{M}, X \models^+ \varphi \quad \text{implies} \quad \mathbb{M}, X \models^+ \varphi \vee \psi$$

because $X = X \cup \varnothing$ and $\mathbb{M}, \varnothing \models^+ \psi$ (even if ψ is tautologically false). If Eloise has a winning strategy for $G(\mathbb{M}, X, \varphi)$, then she can always win the semantic game for $\varphi \vee \psi$ simply by choosing the left disjunct and then executing her winning strategy. Similarly, $\mathbb{M}, X \models^- \varphi$ implies $\mathbb{M}, X \models^- \varphi \wedge \psi$ because if Abelard can falsify φ he can falsify $\varphi \wedge \psi$ by always choosing the left conjunct.

---

[1] Hodges states both properties as Fact 11.1 in [32]. We follow Caicedo *et al.* in calling the former property "downward monotonicity" [9, Lemma 4.6].

If smaller assignment teams correspond to more information about the current assignment, larger assignment teams correspond to less information. In particular, a player can have a winning strategy for $G(\mathbb{M}, X, \varphi)$ and a winning strategy for $G(\mathbb{M}, Y, \varphi)$ without having a winning strategy for $G(\mathbb{M}, X \cup Y, \varphi)$. For instance, in our two-element structure $\mathbb{M} = \{a, b\}$, the IF formula

$$(\exists y/\{x\})x = y$$

is satisfied by $\{s_a\}$ and $\{s_b\}$, but not $\{s_a, s_b\}$.

The fact that the family of winning (or losing) teams for an IF formula in a particular structure is not closed under unions depends in an essential way on the lack of perfect information in the formula's semantic game. We leave it as an exercise for the reader to prove that for every IF formula φ whose slash sets are all empty (i.e., those IF formulas whose semantic games have perfect information) we have $\mathbb{M}, X \models^+ \varphi$ and $\mathbb{M}, Y \models^+ \varphi$ implies $\mathbb{M}, X \cup Y \models^+ \varphi$ (and similarly for $\models^-$).[2]

5.1.1 Substitution

Definition 5.3 Let φ and φ' be IF formulas, let R be an n-ary relation symbol, and let $t_1, \ldots, t_n$ be terms. Let x, y and z be distinct variables, and let W be a finite set of variables such that $x \notin W$. The operation of substituting y for x in a term is defined as for first-order logic (Definition 3.23). Substituting y for x in an IF formula is defined by:

$$
\begin{aligned}
\mathrm{Subst}\big(t_1 = t_2, x, y\big) \quad &\text{is} \quad \mathrm{Subst}(t_1, x, y) = \mathrm{Subst}(t_2, x, y), \\
\mathrm{Subst}\big(R(t_1, \ldots, t_n), x, y\big) \quad &\text{is} \quad R\big(\mathrm{Subst}(t_1, x, y), \\
&\qquad\qquad \ldots, \mathrm{Subst}(t_n, x, y)\big), \\
\mathrm{Subst}(\neg \varphi, x, y) \quad &\text{is} \quad \neg\, \mathrm{Subst}(\varphi, x, y), \\
\mathrm{Subst}(\varphi \circ \varphi', x, y) \quad &\text{is} \quad \mathrm{Subst}(\varphi, x, y) \circ \mathrm{Subst}(\varphi', x, y), \\
\mathrm{Subst}\big((Qx/W)\varphi, x, y\big) \quad &\text{is} \quad (Qx/W)\varphi, \\
\mathrm{Subst}\big((Qz/W)\varphi, x, y\big) \quad &\text{is} \quad (Qz/W)\,\mathrm{Subst}(\varphi, x, y), \\
\mathrm{Subst}\big((Qx/W \cup \{x\})\varphi, x, y\big) \quad &\text{is} \quad (Qx/W \cup \{y\})\varphi, \\
\mathrm{Subst}\big((Qz/W \cup \{x\})\varphi, x, y\big) \quad &\text{is} \quad (Qz/W \cup \{y\})\,\mathrm{Subst}(\varphi, x, y). \quad \dashv
\end{aligned}
$$

---

[2] Saying that the family of winning teams for an IF formula φ is closed under unions is equivalent to saying that an assignment team X satisfies φ if and only if every assignment in X satisfies φ. Hodges calls formulas with the latter property *flat* [33, p. 54].

Observe that we are not allowed to substitute arbitrary terms into IF formulas, only variables. If φ is an IF formula that has a free variable x in a slash set, such as $(\exists y/\{x\})x = y$, the result of substituting a compound term or a constant symbol for x would not be an IF formula. Nor are we allowed to substitute a variable y into the scope of a quantifier (Qy/W), i.e.,

$$\text{Subst}((Qy/W)\varphi, x, y)$$

is not defined. Thus we do not have a full substitution lemma for IF formulas. The next lemma, due to Caicedo *et al.* [9, Lemma 4.13], is as close as we come. For any assignment s, let

$$s_{-x} = s \restriction (\text{dom}(s) - \{x\}).$$

Lemma 5.4 (Interchanging free variables) *Suppose x does not occur bound and y does not occur in φ. For any suitable structure $\mathbb{M}$ and any team X whose domain contains $\text{Free}(\varphi)$, if $x \in \text{dom}(X)$ and $y \notin \text{dom}(X)$, then*

$$\mathbb{M}, X \models^{\pm} \varphi \quad \textit{iff} \quad \mathbb{M}, \left\{ s(y/s(x))_{-x} : s \in X \right\} \models^{\pm} \text{Subst}(\varphi, x, y).$$

Proof On the right-hand side, y takes the place of x in both the team and the formula.   ⊣

5.1.2 Extending assignments

Another difference between first-order logic and IF logic appears when assigning values to additional variables. Extending an assignment does not affect whether it satisfies a first-order formula (assuming the domain of the original assignment contained the free variables of the formula). Extending the assignments in a team can affect whether it satisfies or dissatisfies an IF formula, however, because the extra variables may encode information otherwise unavailable to the players. For example, let $\mathbb{M} = \{a, b\}$, s_a, and s_b be as before, while

$$
\begin{aligned}
s'_{aa} &= \{(x, a), (z, a)\}, & s'_{ba} &= \{(x, b), (z, a)\}, \\
s'_{ab} &= \{(x, a), (z, b)\}, & s'_{bb} &= \{(x, b), (z, b)\}.
\end{aligned}
$$

The formula $(\exists y/\{x\})x = y$ is satisfied by $\{s'_{aa}, s'_{bb}\}$ but not $\{s_a, s_b\}$ because in the first case the value of z signals the value of x.[3] Dually,

---

[3] Is the formula satisfied by $\{s'_{ab}, s'_{ba}\}$?

$(\forall y/\{x\})x \neq y$ is dissatisfied by $\{s'_{aa}, s'_{bb}\}$ but not $\{s_a, s_b\}$. Thus, extending a team of assignments may help Eloise or Abelard depending on the situation.

One way to avoid encoding illicit information in the new variables is to extend every assignment in the same way. That way the players cannot infer anything about the original assignment from the values of the new variables. For example, the formula $(\exists y/\{x\})x = y$ is not satisfied by $\{s'_{aa}, s'_{ba}\}$, $\{s'_{ab}, s'_{bb}\}$, or $\{s'_{aa}, s'_{ab}, s'_{ba}, s'_{bb}\}$ because for any $\{x\}$-uniform function f we must have $f(s'_{aa}) = f(s'_{ba})$ and $f(s'_{ab}) = f(s'_{bb})$. The general principle that uniformly extending every assignment in a team does not encode any additional information by the team is formalized by the following theorem, which is [9, Theorem 5.1].

Theorem 5.5 (Cartesian extension) *Let φ be an IF formula, $\mathbb{M}$ a suitable structure, and X a team of assignments whose domain contains Free(φ). For any nonempty assignment team Y whose domain is disjoint from that of X,*

$$\mathbb{M}, X \models^{\pm} \varphi \quad \text{iff} \quad \mathbb{M}, X \times Y \models^{\pm} \varphi.$$

Proof If φ is a literal, the theorem follows immediately from Proposition 3.13. Since $\models^-$ is defined dually to $\models^+$ it suffices to prove the cases for $\models^+$ in the inductive step.

Suppose φ is $\psi_1 \vee \psi_2$. If $\mathbb{M}, X \models^+ \psi_1 \vee \psi_2$ there exists a cover $X_1 \cup X_2 = X$ such that $\mathbb{M}, X_1 \models^+ \psi_1$ and $\mathbb{M}, X_2 \models^+ \psi_2$. By inductive hypothesis

$$\mathbb{M}, X_1 \times Y \models^+ \psi_1 \quad \text{and} \quad \mathbb{M}, X_2 \times Y \models^+ \psi_2,$$

from which it follows that $\mathbb{M}, X \times Y \models^+ \psi_1 \vee \psi_2$. Conversely, by downward monotonicity (Proposition 5.1), the previous statement implies that for any $s \in Y$ we have $\mathbb{M}, X \times \{s\} \models^+ \psi_1 \vee \psi_2$. Hence for some cover $(X_1 \times \{s\}) \cup (X_2 \times \{s\}) = X \times \{s\}$ we have

$$\mathbb{M}, X_1 \times \{s\} \models^+ \psi_1 \quad \text{and} \quad \mathbb{M}, X_2 \times \{s\} \models^+ \psi_2.$$

Then by inductive hypothesis $\mathbb{M}, X_1 \models^+ \psi_1$ and $\mathbb{M}, X_2 \models^+ \psi_2$, which implies $\mathbb{M}, X \models^+ \psi_1 \vee \psi_2$.

Suppose φ is $\psi_1 \wedge \psi_2$. By inductive hypothesis,

$$
\begin{aligned}
\mathbb{M}, X \models^+ \psi_1 \wedge \psi_2 \quad &\text{iff} \quad \mathbb{M}, X \models^+ \psi_1 \text{ and } \mathbb{M}, X \models^+ \psi_2 \\
&\text{iff} \quad \mathbb{M}, X \times Y \models^+ \psi_1 \text{ and } \mathbb{M}, X \times Y \models^+ \psi_2 \\
&\text{iff} \quad \mathbb{M}, X \times Y \models^+ \psi_1 \wedge \psi_2.
\end{aligned}
$$

Suppose φ is $(\exists x/W)\psi$. If $\mathbb{M}, X \models^+ (\exists x/W)\psi$ then there exists a W-uniform function $f\colon X \to M$ such that $\mathbb{M}, X[x, f] \models^+ \psi$. By inductive hypothesis $\mathbb{M}, X[x, f] \times Y' \models^+ \psi$, where $Y' = \{\, s_{-x} : s \in Y \,\}$. Observe that

$$X[x, f] \times Y' = (X \times Y)[x, g]$$

where $g\colon X \times Y \to M$ is the W-uniform function defined by

$$g(s) = f\big(s \restriction \mathrm{dom}(X)\big).$$

Thus $\mathbb{M}, X \times Y \models^+ (\exists x/W)\psi$. Conversely, if there exists a W-uniform function $g\colon X \times Y \to M$ such that $\mathbb{M}, (X \times Y)[x, g] \models^+ \psi$, then for any $s_0 \in Y$,

$$\mathbb{M}, \big(X \times \{s_0\}\big)[x, g] \models^+ \psi$$

by downward monotonicity. Let $f\colon X \to M$ be the W-uniform function defined by $f(s) = g\big(s \cup s_0\big)$. Then

$$\mathbb{M}, X[x, f] \times \big\{(s_0)_{-x}\big\} \models^+ \psi,$$

so by inductive hypothesis $\mathbb{M}, X[x, f] \models^+ \psi$. Thus $\mathbb{M}, X \models^+ (\exists x/W)\psi$.

Suppose φ is $(\forall x/W)\psi$. Then by inductive hypothesis

$$
\begin{array}{rll}
\mathbb{M}, X \models^+ (\forall x/W)\psi & \text{iff} & \mathbb{M}, X[x, M] \models^+ \psi \\
& \text{iff} & \mathbb{M}, X[x, M] \times Y' \models^+ \psi \\
& \text{iff} & \mathbb{M}, (X \times Y)[x, M] \models^+ \psi \\
& \text{iff} & \mathbb{M}, X \times Y \models^+ (\forall x/W)\psi,
\end{array}
$$

where $Y' = \{s_{-x} : x \in Y\}$. $\qquad\dashv$

Extending a winning team cannot hurt Eloise's chances of winning — nor can extending a losing team hurt Abelard — because the winning player can simply ignore the information stored in the extra variables.

Theorem 5.6 (Caicedo *et al.* [9, Theorem 5.3]) *Let φ be an IF formula, $\mathbb{M}$ a suitable structure, and X a team of assignments whose domain contains* $\mathrm{Free}(\varphi)$. *If V is a finite set of variables disjoint from* $\mathrm{dom}(X)$, *and Y is an extension of X to* $\mathrm{dom}(X) \cup V$,

$$\mathbb{M}, X \models^\pm \varphi \quad \text{implies} \quad \mathbb{M}, Y \models^\pm \varphi.$$

Proof By the Cartesian extension theorem (Theorem 5.5) $\mathbb{M}, X \models^\pm \varphi$ implies $\mathbb{M}, X \times M^V \models^\pm \varphi$, which by downward monotonicity implies $\mathbb{M}, Y \models^\pm \varphi$. $\qquad\dashv$

The converse of Theorem 5.6 does not hold because the variables in V may encode information that enables the players to follow otherwise forbidden strategies (see pp. 90–91). However, we can prevent the players from using any additional information encoded by the extended team simply by adding the extra variables to every slash set in the formula.

Definition 5.7 Let φ be an IF formula, and let V be a finite set of variables. The formula φ/V is obtained by making every quantifier independent of V:

$$\varphi/V = \varphi \quad (\varphi \text{ literal}),$$
$$(\varphi \circ \varphi')/V = (\varphi/V) \circ (\varphi'/V),$$
$$[(Qx/W)\varphi]/V = (Qx/W \cup V)[\varphi/V]. \qquad \dashv$$

Lemma 5.8 (Caicedo *et al.* [9, Lemma 5.5]) *Let φ be an IF formula, $\mathbb{M}$ a suitable structure, and X a team of assignments whose domain contains* Free(φ). *If V is a finite set of variables that occur neither in φ nor in* dom(X), *and Y is an extension of X to* dom$(X) \cup V$,

$$\mathbb{M}, X \models^{\pm} \varphi \quad \text{iff} \quad \mathbb{M}, Y \models^{\pm} \varphi/V.$$

Proof If φ is a literal, then φ/V is simply φ, and the theorem follows from Proposition 3.13.

Suppose φ is $\psi_1 \vee \psi_2$. Then $\mathbb{M}, X \models^{+} \psi_1 \vee \psi_2$ implies $\mathbb{M}, X_1 \models^{+} \psi_1$ and $\mathbb{M}, X_2 \models^{+} \psi_2$ for some cover $X_1 \cup X_2 = X$. Let

$$Y_1 = \left\{ s \in Y : s \restriction \text{dom}(X) \in X_1 \right\},$$
$$Y_2 = \left\{ s \in Y : s \restriction \text{dom}(X) \in X_2 \right\}.$$

Then Y_i is an extension of X_i to dom$(X_i) \cup V$, so by inductive hypothesis

$$\mathbb{M}, Y_1 \models^{+} \psi_1/V \quad \text{and} \quad \mathbb{M}, Y_2 \models^{+} \psi_2/V.$$

Since $Y_1 \cup Y_2 = Y$ we have $\mathbb{M}, Y \models^{+} (\psi_1 \vee \psi_2)/V$. Conversely, suppose

$$\mathbb{M}, Y_1 \models^{+} \psi_1/V \quad \text{and} \quad \mathbb{M}, Y_2 \models^{+} \psi_2/V$$

for some cover $Y_1 \cup Y_2 = Y$. Let $X_i = \left\{ s \restriction \text{dom}(X) : s \in Y_i \right\}$. Then Y_i is an extension of X_i to dom$(X) \cup V$, so by inductive hypothesis

$$\mathbb{M}, X_1 \models^{+} \psi_1 \quad \text{and} \quad \mathbb{M}, X_2 \models^{+} \psi_2.$$

Since $X_1 \cup X_2 = X$ we have $\mathbb{M}, X \models^{+} \psi_1 \vee \psi_2$.

Suppose φ is $\psi_1 \wedge \psi_2$. Then by inductive hypothesis

$$
\begin{aligned}
\mathbb{M}, X \models^+ \psi_1 \wedge \psi_2 \quad &\text{iff} \quad \mathbb{M}, X \models^+ \psi_1 \text{ and } \mathbb{M}, X \models^+ \psi_2 \\
&\text{iff} \quad \mathbb{M}, Y \models^+ \psi_1/V \text{ and } \mathbb{M}, Y \models^+ \psi_2/V \\
&\text{iff} \quad \mathbb{M}, Y \models^+ (\psi_1 \wedge \psi_2)/V.
\end{aligned}
$$

Suppose φ is $(\exists x/W)\psi$. If $\mathbb{M}, X \models^+ (\exists x/W)\psi$ there is a W-uniform function $f \colon X \to M$ such that $\mathbb{M}, X[x, f] \models^+ \psi$. Define a function $g \colon Y \to M$ by

$$
g(s) = f\big(s \restriction \mathrm{dom}(X)\big).
$$

Then $Y[x, g]$ is an extension of $X[x, f]$ to $\mathrm{dom}\big(X[x, f]\big) \cup V$, so by inductive hypothesis $\mathbb{M}, Y[x, g] \models^+ \psi/V$. Since g is uniform in $W \cup V$ it follows that $\mathbb{M}, Y \models^+ (\exists x/W \cup V)\psi/V$. Conversely, suppose there is a $W \cup V$-uniform function $g \colon Y \to M$ such that $\mathbb{M}, Y[x, g] \models^+ \psi/V$. Then there is a W-uniform function $f \colon X \to M$ defined by

$$
f\big(s \restriction \mathrm{dom}(X)\big) = g(s)
$$

such that $Y[x, g]$ is an extension of $X[x, f]$ to $\mathrm{dom}\big(X[x, f]\big) \cup V$, so by inductive hypothesis $\mathbb{M}, X[x, f] \models^+ \psi$. Thus $\mathbb{M}, X \models^+ (\exists x/W)\psi$.

Finally, suppose φ is $(\forall x/W)\psi$. Then by inductive hypothesis

$$
\begin{aligned}
\mathbb{M}, X \models^+ (\forall x/W)\psi \quad &\text{iff} \quad \mathbb{M}, X[x, M] \models^+ \psi \\
&\text{iff} \quad \mathbb{M}, Y[x, M] \models^+ \psi/V \\
&\text{iff} \quad \mathbb{M}, Y \models^+ (\forall x/W \cup V)\psi/V
\end{aligned}
$$

because $Y[x, M]$ is an extension of $X[x, M]$ to $\mathrm{dom}\big(X[x, M]\big) \cup V$. $\quad \dashv$

Given all the complications due to signaling we have encountered thus far, we should be extremely careful when evaluating IF formulas. The reader may reasonably wonder whether IF sentences are susceptible to signaling. Fortunately, they are not. Any information stored in a variable before the semantic game for an IF sentence begins cannot signal any useful information to the players.

Theorem 5.9 (Caicedo *et al.* [9, Theorem 5.2]) *Let φ be an IF sentence, $\mathbb{M}$ a suitable structure, and X a nonempty team of assignments in $\mathbb{M}$. Then $\mathbb{M} \models^{\pm} \varphi$ if and only if $\mathbb{M}, X \models^{\pm} \varphi$.*

Proof By the Cartesian extension theorem,

$$\mathbb{M} \models^{\pm} \varphi \quad \text{iff} \quad \mathbb{M}, \{\varnothing\} \models^{\pm} \varphi$$
$$\text{iff} \quad \mathbb{M}, \{\varnothing\} \times X \models^{\pm} \varphi$$
$$\text{iff} \quad \mathbb{M}, X \models^{\pm} \varphi. \qquad \dashv$$

5.2 Extensions of IF logic

In this section we consider two extensions of IF logic that are defined in terms of the basic syntax presented in the last chapter.

5.2.1 Negation

To simplify the presentation of the semantics for IF logic, we only allowed the negation symbol $\neg$ to appear directly in front of atomic formulas. We now relax that restriction by defining $\neg\varphi$ as the dual of φ. When φ is atomic, we have already defined $\neg(\neg\varphi)$ to be φ. We now extend the definition to compound IF formulas:

$$\neg(\varphi \vee \varphi') \quad \text{is} \quad \neg\varphi \wedge \neg\varphi',$$
$$\neg(\varphi \wedge \varphi') \quad \text{is} \quad \neg\varphi \vee \neg\varphi',$$
$$\neg(\exists x/W)\varphi \quad \text{is} \quad (\forall x/W)\neg\varphi,$$
$$\neg(\forall x/W)\varphi \quad \text{is} \quad (\exists x/W)\neg\varphi.$$

Lemma 5.10 *Let φ be an IF formula, $\mathbb{M}$ a suitable structure, and X a team of assignments whose domain contains Free(φ). Then*

$$\mathbb{M}, X \models^{\pm} \neg\varphi \quad \textit{iff} \quad \mathbb{M}, X \models^{\mp} \varphi.$$

Proof If φ is a literal, then

$$\mathbb{M}, X \models^{+} \neg\varphi \quad \text{iff} \quad \mathbb{M}, s \models \neg\varphi \quad \text{(for all } s \in X\text{)}$$
$$\text{iff} \quad \mathbb{M}, s \not\models \varphi \quad \text{(for all } s \in X\text{)}$$
$$\text{iff} \quad \mathbb{M}, X \models^{-} \varphi,$$

$$\mathbb{M}, X \models^{-} \neg\varphi \quad \text{iff} \quad \mathbb{M}, s \not\models \neg\varphi \quad \text{(for all } s \in X\text{)}$$
$$\text{iff} \quad \mathbb{M}, s \models \varphi \quad \text{(for all } s \in X\text{)}$$
$$\text{iff} \quad \mathbb{M}, X \models^{+} \varphi.$$

Suppose φ is $\psi \vee \psi'$. Then by inductive hypothesis,

$$
\begin{aligned}
\mathbb{M}, X \models^+ \neg(\psi \vee \psi') \quad &\text{iff} \quad \mathbb{M}, X \models^+ \neg\psi \wedge \neg\psi' \\
&\text{iff} \quad \mathbb{M}, X \models^+ \neg\psi \text{ and } \mathbb{M}, X \models^+ \neg\psi' \\
&\text{iff} \quad \mathbb{M}, X \models^- \psi \text{ and } \mathbb{M}, X \models^- \psi' \\
&\text{iff} \quad \mathbb{M}, X \models^- \psi \vee \psi'.
\end{aligned}
$$

Suppose φ is $\psi \wedge \psi'$. Then by inductive hypothesis $\mathbb{M}, X \models^+ \neg(\psi \wedge \psi')$ if and only if $\mathbb{M}, X \models^+ \neg\psi \vee \neg\psi'$ if and only if there exists a cover $Y \cup Y' = X$ such that $\mathbb{M}, Y \models^+ \neg\psi$ and $\mathbb{M}, Y' \models^+ \neg\psi'$ if and only if there exists a cover $Y \cup Y' = X$ such that $\mathbb{M}, Y \models^- \psi$ and $\mathbb{M}, Y' \models^- \psi'$ if and only if $\mathbb{M}, X \models^- \psi \wedge \psi'$.

Suppose φ is $(\exists x/W)\psi$. Then by inductive hypothesis

$$
\begin{aligned}
\mathbb{M}, X \models^+ \neg(\exists x/W)\psi \quad &\text{iff} \quad \mathbb{M}, X \models^+ (\forall x/W)\neg\psi \\
&\text{iff} \quad \mathbb{M}, X[x, M] \models^+ \neg\psi \\
&\text{iff} \quad \mathbb{M}, X[x, M] \models^- \psi \\
&\text{iff} \quad \mathbb{M}, X \models^- (\exists x/W)\psi.
\end{aligned}
$$

Suppose φ is $(\forall x/W)\psi$. Then by inductive hypothesis

$$
\mathbb{M}, X \models^+ \neg(\forall x/W)\psi
$$

if and only if $\mathbb{M}, X \models^+ (\exists x/W)\neg\psi$ if and only if there is a W-uniform $f \colon X \to M$ such that $\mathbb{M}, X[x, f] \models^+ \neg\psi$ if and only if there is a W-uniform $f \colon X \to M$ such that $\mathbb{M}, X[x, f] \models^- \psi$ if and only if $\mathbb{M}, X \models^- (\forall x/W)\psi$. $\dashv$

5.2.2 Slashed connectives

Connectives prompt Eloise or Abelard to choose a disjunct or a conjunct, respectively. A player's choice may depend on the values already given to certain variables, so we can extend IF logic by attaching slash sets to connectives. If we had included the slashed connectives $\vee_{/W}$ and $\wedge_{/W}$ in the basic syntax for IF logic, the indistinguishability relation for connectives on page 63 would have read:

- For $h, h' \in H_{\chi_1 \vee_{/W} \chi_2}$ we have $h \sim_\exists h'$ if and only if $s_h \approx_W s_{h'}$.
  For $h, h' \in H_{\chi_1 \wedge_{/W} \chi_2}$ we have $h \sim_\forall h'$ if and only if $s_h \approx_W s_{h'}$.

Thus in the semantic game for $\varphi \vee_{/W} \psi$ Eloise must choose φ or ψ without knowing the values of the variables in W. Abelard faces similar restrictions in the semantic game for $\varphi \wedge_{/W} \psi$.

We did not include slashed connectives in the basic syntax of IF logic for two reasons. First, we wished to keep the basic syntax of IF logic as simple as possible; second, we can simulate slashed connectives using slashed quantifiers. The trick is to allow Eloise (Abelard) to store her (his) choice of disjunct (conjunct) in a special-purpose variable that plays no other role in the formula.

From now on, let $\varphi \vee_{/W} \psi$ be an abbreviation for

$$(\exists z/W)\Big[\big(z = 0 \wedge \varphi/\{z\}\big) \vee \big(z = 1 \wedge \psi/\{z\}\big)\Big],$$

where 0 and 1 are constant symbols, and z is a fresh variable that does not occur in φ, ψ, or W. Eloise uses the value of z to declare which disjunct she will choose, the conjuncts $z = 0$ and $z = 1$ force her to stick to her word, and the slash set in $(\exists z/W)$ prevents her from seeing the variables in W when making her choice. Similarly, we can restrict the information available to Abelard by taking $\varphi \wedge_{/W} \psi$ to be an abbreviation for $\neg(\neg\varphi \vee_{/W} \neg\psi)$.

In order to interpret IF formulas with slashed connectives compositionally, we need a way to split a team of assignments X into two subteams Y and Y' such that the subteam to which a given assignment belongs does not depend on the values of variables in W.

The following definition is a modified version of [9, Definition 4.1].

Definition 5.11 Let X be a team of assignments with $W \subseteq \mathrm{dom}(X)$. A subteam $Y \subseteq X$ is W-*saturated* in X if it is closed under $\approx_W$, i.e., for all $s, s' \in X$,

$$s \approx_W s' \text{ and } s' \in Y \quad \text{imply} \quad s \in Y.$$

A cover $Y \cup Y' = X$ is *uniform in* W (or W-*uniform*) if both Y and Y' are W-saturated in X. ⊣

If Y and Y' are W-saturated in X, then so are $Y \cup Y'$, $Y \cap Y'$, and $X - Y$. Hence, whenever $Y \cup Y' = X$ is a W-uniform cover, we may assume Y and Y' are disjoint.

Lemma 5.12 *Let X be a team of assignments with domain U.*

(a) Every cover $Y \cup Y' = X$ is uniform in $\varnothing$.
(b) A cover $Y \cup Y' = X$ is uniform in U if and only if $Y = X$ or $Y' = X$.

Proof (a) Suppose $Y \cup Y' = X$. If $s \approx_\emptyset s'$ then $s = s'$, so both Y and Y' are $\emptyset$-saturated in X. (b) For all $s, s' \in X$ we have $s \approx_U s'$, so Y is U-saturated in X if and only if $Y = \emptyset$ or $Y = X$.   ⊣

Theorem 5.13 *Let φ and ψ be IF$_L$ formulas, $\mathbb{M}$ an L-structure, and $\mathbb{M}^*$ an expansion of $\mathbb{M}$ to the vocabulary $L \cup \{0, 1\}$ in which the constant symbols 0 and 1 are interpreted by distinct elements, if possible. Let X be a team of assignments whose domain contains* $\mathrm{Free}(\varphi)$, $\mathrm{Free}(\psi)$, *and* W. *Finally, let z be a variable that does not occur in φ, ψ, or* $\mathrm{dom}(X)$. *Then on the truth axis,*

$$\mathbb{M}^*, X \models^+ (\exists z/W)\Big[(z = 0 \wedge \varphi/\{z\}) \vee (z = 1 \wedge \psi/\{z\})\Big]$$

if and only if there exists a W-uniform cover $Y \cup Y' = X$ such that

$$\mathbb{M}, Y \models^+ \varphi \quad and \quad \mathbb{M}, Y' \models^+ \psi.$$

On the falsity axis,

$$\mathbb{M}^*, X \models^- (\exists z/W)\Big[(z = 0 \wedge \varphi/\{z\}) \vee (z = 1 \wedge \psi/\{z\})\Big]$$

if and only if $\mathbb{M}, X \models^- \varphi$ *and* $\mathbb{M}, X \models^- \psi$.

Proof If $|M| = 1$ the theorem is trivial, so assume that $|M| \geq 2$ and $0^{\mathbb{M}^*} \neq 1^{\mathbb{M}^*}$. Starting with the reverse implication, suppose $\mathbb{M}, Y \models^+ \varphi$ and $\mathbb{M}, Y' \models^+ \psi$ for some W-uniform cover $Y \cup Y' = X$. We may assume without loss of generality that $Y \cap Y' = \emptyset$. Define a W-uniform function $f \colon X \to M$ by

$$f(s) = \begin{cases} 0^{\mathbb{M}^*} & \text{if } s \in Y, \\ 1^{\mathbb{M}^*} & \text{if } s \in Y'. \end{cases}$$

Then by Lemma 5.8

$$\mathbb{M}^*, Y[z, 0^{\mathbb{M}^*}] \models^+ z = 0 \wedge \varphi/\{z\},$$
$$\mathbb{M}^*, Y'[z, 1^{\mathbb{M}^*}] \models^+ z = 1 \wedge \psi/\{z\},$$

and since $Y[z, 0^{\mathbb{M}^*}] \cup Y'[z, 1^{\mathbb{M}^*}] = X[z, f]$ we have

$$\mathbb{M}^*, X[z, f] \models^+ (z = 0 \wedge \varphi/\{z\}) \vee (z = 1 \wedge \psi/\{z\}),$$

which implies $\mathbb{M}^*, X \models^+ (\exists z/W)\Big[(z = 0 \wedge \varphi/\{z\}) \vee (z = 1 \wedge \psi/\{z\})\Big]$. Conversely, suppose there is a W-uniform function $f \colon X \to M$ such that

$$\mathbb{M}^*, X[z, f] \models^+ (z = 0 \wedge \varphi/\{z\}) \vee (z = 1 \wedge \psi/\{z\}).$$

Then

$$\mathbb{M}^*, Y\left[z, 0^{\mathbb{M}^*}\right] \models^+ z = 0 \wedge \varphi/\{z\},$$
$$\mathbb{M}^*, Y'\left[z, 1^{\mathbb{M}^*}\right] \models^+ z = 1 \wedge \psi/\{z\},$$

for some cover $Y \cup Y' = X$ which is uniform in W because f is. Since z does not occur in φ or ψ, Lemma 5.8 tells us that

$$\mathbb{M}^*, Y \models^+ \varphi \quad \text{and} \quad \mathbb{M}^*, Y' \models^+ \psi$$

which implies $\mathbb{M}, Y \models^+ \varphi$ and $\mathbb{M}, Y' \models^+ \psi$.

On the falsity axis, it suffices to show that

$$\mathbb{M}^*, X \models^+ (\forall z/W)\left[(z \neq 0 \vee \neg\varphi/\{z\}) \wedge (z \neq 1 \vee \neg\psi/\{z\})\right]$$

if and only if $\mathbb{M}, X \models^+ \neg\varphi$ and $\mathbb{M}, X \models^+ \neg\psi$. Suppose the latter. Then by Lemma 5.8 we have

$$\mathbb{M}^*, X[z, M] \models^+ \neg\varphi/\{z\} \quad \text{and} \quad \mathbb{M}^*, X[z, M] \models^+ \neg\psi/\{z\}.$$

Thus $\mathbb{M}^*, X[z, M] \models^+ (z \neq 0 \vee \neg\varphi/\{z\}) \wedge (z \neq 1 \vee \neg\psi/\{z\})$ because $z \neq 0$ and $z \neq 1$ are both satisfied by the empty team of assignments. Therefore

$$\mathbb{M}^*, X \models^+ (\forall z/W)\left[(z \neq 0 \vee \neg\varphi/\{z\}) \wedge (z \neq 1 \vee \neg\psi/\{z\})\right].$$

Conversely, suppose

$$\mathbb{M}^*, X[z, M] \models^+ (z \neq 0 \vee \neg\varphi/\{z\}) \wedge (z \neq 1 \vee \neg\psi/\{z\}).$$

Then

$$\mathbb{M}^*, X[z, M] \models^+ z \neq 0 \vee \neg\varphi/\{z\},$$
$$\mathbb{M}^*, X[z, M] \models^+ z \neq 1 \vee \neg\psi/\{z\},$$

which implies $\mathbb{M}^*, X\left[z, 0^{\mathbb{M}^*}\right] \models^+ \neg\varphi/\{z\}$ and $\mathbb{M}^*, X\left[z, 1^{\mathbb{M}^*}\right] \models^+ \neg\psi/\{z\}$, so by Lemma 5.8 we have $\mathbb{M}, X \models^+ \neg\varphi$ and $\mathbb{M}, X \models^+ \neg\psi$. ⊣

Dually we have $\mathbb{M}^*, X \models^+ \varphi \wedge_{/W} \psi$ if and only if $\mathbb{M}, X \models^+ \varphi$ and $\mathbb{M}, X \models^+ \psi$, while $\mathbb{M}^*, X \models^- \varphi \wedge_{/W} \psi$ if and only if there exists a W-uniform cover $Y \cup Y' = X$ such that $\mathbb{M}, Y \models^- \varphi$ and $\mathbb{M}, Y' \models^- \psi$.

5.3 Logical equivalence

One IF sentence entails another if whenever Eloise can verify the former she can also verify the latter, and whenever Abelard can falsify the latter he can falsify the former. Two IF sentences are logically equivalent if they entail each other. Checking whether two IF sentences are equivalent is a more complicated affair than verifying the equivalence of two first-order sentences because we must keep track of truth and falsity separately.

When checking the equivalence of IF formulas with free variables, we must also take into account the context of the formula. Any variable that has been assigned a value can affect the strategies of the players, even when the variable does not occur in the formula. Thus we will only consider the logical equivalence of IF formulas relative to a specified set of variables.

Throughout this section φ, ψ, and χ will be IF formulas, and $W, V \subseteq U$ will be finite sets of variables such that U contains $\mathrm{Free}(\varphi)$, $\mathrm{Free}(\psi)$, and $\mathrm{Free}(\chi)$.

Definition 5.14   We write $\varphi \models_U^+ \psi$, and say φ *truth entails* ψ *relative to* U, if for every suitable structure $\mathbb{M}$ and assignment team X with domain U,

$$\mathbb{M}, X \models^+ \varphi \quad \text{implies} \quad \mathbb{M}, X \models^+ \psi.$$

Similarly, we write $\varphi \models_U^- \psi$, and say φ *falsity entails* ψ *relative to* U, if for every suitable structure $\mathbb{M}$ and assignment team X with domain U,

$$\mathbb{M}, X \models^- \varphi \quad \text{implies} \quad \mathbb{M}, X \models^- \psi.$$

We write $\varphi \models_U \psi$, and say φ *entails* ψ *relative to* U, if

$$\varphi \models_U^+ \psi \quad \text{and} \quad \psi \models_U^- \varphi.$$

We write $\varphi \equiv_U^+ \psi$, and say φ is *truth equivalent to* ψ *relative to* U, if

$$\varphi \models_U^+ \psi \quad \text{and} \quad \psi \models_U^+ \varphi.$$

We write $\varphi \equiv_U^- \psi$, and say φ is *falsity equivalent to* ψ *relative to* U, if

$$\varphi \models_U^- \psi \quad \text{and} \quad \psi \models_U^- \varphi.$$

We write $\varphi \equiv_U \psi$, and say φ is *equivalent to* ψ *relative to* U, if

$$\varphi \models_U \psi \quad \text{and} \quad \psi \models_U \varphi.$$

As usual, we omit the subscript U when it is empty. Thus, when φ and ψ are IF sentences we may simply write

$$\varphi \models^\pm \psi, \quad \varphi \models \psi, \quad \varphi \equiv^\pm \psi, \quad \varphi \equiv \psi,$$

and say φ *truth/falsity entails* ψ, φ *entails* ψ, φ is *truth/falsity equivalent to* ψ, and φ is *equivalent* to ψ, respectively.   ⊣

The following example shows that the domain of the assignment teams can affect whether two IF formulas are equivalent.

Example 5.15 We wish to show that

$$(\exists y/\{x\})x = y \equiv_{\{x\}} \forall z(\exists y/\{x\})x = y,$$
$$(\exists y/\{x\})x = y \not\equiv_{\{x,z\}} \forall z(\exists y/\{x\})x = y.$$

Suppose $\mathbb{M}$ is a structure and X is a team of assignments with domain $\{x\}$. Then $X[z, M] = X \times M^{\{z\}}$ is a Cartesian extension of X, so by the Cartesian extension theorem

$$\mathbb{M}, X \models^+ (\exists y/\{x\})x = y \quad \text{iff} \quad \mathbb{M}, X[z, M] \models^+ (\exists y/\{x\})x = y$$
$$\text{iff} \quad \mathbb{M}, X \models^+ \forall z(\exists y/\{x\})x = y.$$

Thus the two formulas are truth equivalent relative to $\{x\}$. They are falsity equivalent relative to $\{x\}$ because no team with domain $\{x\}$ dissatisfies either formula.

Let $\mathbb{M} = \{a, b\}$ be a two-element structure, and let

$$s_a = \{(x, a)\}, \qquad\qquad s_b = \{(x, b)\},$$
$$s'_{aa} = \{(x, a), (z, a)\}, \qquad s'_{ba} = \{(x, b), (z, a)\},$$
$$s'_{ab} = \{(x, a), (z, b)\}, \qquad s'_{bb} = \{(x, b), (z, b)\}.$$

Then $\mathbb{M}, \{s'_{aa}, s'_{bb}\} \models^+ (\exists y/\{x\})x = y$ (see Example 4.26), but by the discussion on page 91,

$$\mathbb{M}, \{s'_{aa}, s'_{bb}\} \not\models^+ \forall z(\exists y/\{x\})x = y.$$

Hence the two formulas are not truth equivalent relative to $\{x, z\}$. ⊣

Negation reverses the order of entailment between two IF formulas, while connectives and quantifiers preserve it.

Proposition 5.16 $\varphi \models^{\pm}_U \psi$ *if and only if* $\neg\varphi \models^{\mp}_U \neg\psi$. *Thus*

$$\varphi \models_U \psi \quad \text{iff} \quad \neg\psi \models_U \neg\varphi.$$

Proof Let $\mathbb{M}$ be a suitable structure, and let X be a team of assignments with domain U. If $\varphi \models^+_U \psi$ then

$$\mathbb{M}, X \models^- \neg\varphi \quad \text{implies} \quad \mathbb{M}, X \models^+ \varphi$$
$$\text{implies} \quad \mathbb{M}, X \models^+ \psi$$
$$\text{implies} \quad \mathbb{M}, X \models^- \neg\psi.$$

Likewise, if $\varphi \models_U^- \psi$ then

$$\begin{aligned}
\mathbb{M}, X \models^+ \neg\varphi \quad &\text{implies} \quad \mathbb{M}, X \models^- \varphi \\
&\text{implies} \quad \mathbb{M}, X \models^- \psi \\
&\text{implies} \quad \mathbb{M}, X \models^+ \neg\psi.
\end{aligned}$$

The converse is symmetrical. Thus $\varphi \models_U^+ \psi$ and $\psi \models_U^- \varphi$ if and only if $\neg\varphi \models_U^- \neg\psi$ and $\neg\psi \models_U^+ \neg\varphi$. ⊣

Proposition 5.17 (a) *Suppose* $\varphi \models_U \varphi'$ *and* $\psi \models_U \psi'$. *Then*

$$\varphi \vee_{/W} \psi \models_U \varphi' \vee_{/W} \psi',$$
$$\varphi \wedge_{/W} \psi \models_U \varphi' \wedge_{/W} \psi'.$$

(b) *Suppose* $\varphi \models_{U \cup \{x\}} \psi$. *Then*

$$(\exists x/W)\varphi \models_U (\exists x/W)\psi,$$
$$(\forall x/W)\varphi \models_U (\forall x/W)\psi.$$

Proof Let $\mathbb{M}$ be a suitable structure, and let X be a team of assignments in $\mathbb{M}$ with domain U.

(a) If $\mathbb{M}, X \models^+ \varphi \vee_{/W} \psi$ then there is a W-uniform cover $X = Y \cup Z$ such that $\mathbb{M}, Y \models^+ \varphi$ and $\mathbb{M}, Z \models^+ \psi$ which implies by hypothesis that $\mathbb{M}, Y \models^+ \varphi'$ and $\mathbb{M}, Z \models^+ \psi'$. Hence $\mathbb{M}, X \models^+ \varphi' \vee_{/W} \psi'$. On the falsity axis, if $\mathbb{M}, X \models^- \varphi' \vee_{/W} \psi'$, then $\mathbb{M}, X \models^- \varphi'$ and $\mathbb{M}, X \models^- \psi'$, which implies $\mathbb{M}, X \models^- \varphi$ and $\mathbb{M}, X \models^- \psi$. Hence $\mathbb{M}, X \models^- \varphi \vee_{/W} \psi$.

By Proposition 5.16, the hypotheses imply that $\neg\varphi' \models_U \neg\varphi$ and $\neg\psi' \models_U \neg\psi$, which by the previous argument implies

$$\neg\varphi' \vee_{/W} \neg\psi' \models_U \neg\varphi \vee_{/W} \neg\psi.$$

Hence $\varphi \wedge_{/W} \psi \models_U \varphi' \wedge_{/W} \psi'$.

(b) If $\mathbb{M}, X \models^+ (\exists x/W)\varphi$ there is a W-uniform function $f\colon X \to M$ such that $\mathbb{M}, X[x, f] \models^+ \varphi$, which implies by hypothesis that

$$\mathbb{M}, X[x, f] \models^+ \psi.$$

Hence $\mathbb{M}, X \models^+ (\exists x/W)\psi$. On the falsity axis, if $\mathbb{M}, X \models^- (\exists x/W)\psi$ then $\mathbb{M}, X[x, M] \models^- \psi$, which by hypothesis implies $\mathbb{M}, X[x, M] \models^- \varphi$. Hence $\mathbb{M}, X \models^- (\exists x/W)\varphi$. The dual follows from Proposition 5.16. ⊣

Proposition 5.17 tells us how to substitute equivalent subformulas into an IF sentence to obtain an equivalent IF sentence. For example, if $\varphi \equiv_{\{x,y\}} \psi$ then

$$\forall x (\exists y/\{x\})[\varphi \wedge \chi] \equiv \forall x (\exists y/\{x\})[\psi \wedge \chi].$$

We can also substitute equivalent subformulas into open IF formulas, as long as we keep track of the variables that have been assigned a value.

Theorem 5.18 *Suppose ψ is a subformula of the IF formula φ, and let φ' be the result of replacing ψ with another formula ψ'. If $x_1, \ldots, x_n$ are the variables quantified superordinate to ψ in φ, and $\psi \equiv^{\pm}_{U \cup \{x_1, \ldots, x_n\}} \psi'$, then $\varphi \equiv^{\pm}_{U} \varphi'$.* ⊣

To see why it is necessary for ψ and ψ' to be equivalent relative to $U \cup \{x_1, \ldots, x_n\}$ and not just U, consider the IF formulas

$$(\exists y/\{x\})x = y \quad \text{and} \quad \forall z(\exists y/\{x\})x = y.$$

In both, x is the only free variable. Furthermore, in Example 5.15 we saw that the two formulas are equivalent relative to $\{x\}$, but they are not equivalent relative to $\{x, z\}$. Thus substituting one for the other within the scope of $\forall x \exists z$, for example, may alter the meaning of the larger formula. For instance, the IF sentences

$$\forall x \exists z(\exists y/\{x\})x = y,$$
$$\forall x \exists z \forall z(\exists y/\{x\})x = y,$$

are not equivalent. The former is the signaling sentence, which is true in every structure, whereas the second is not true in any structure with at least two elements because Abelard gets to overwrite the value Eloise assigns to z, thus blocking the signal.

5.3.1 Duality

Logical equivalences (and entailments) between IF formulas always come in pairs. When we wish to prove a pair of equivalences such as

$$\varphi \vee_{/W} \psi \equiv_U \psi \vee_{/W} \varphi,$$
$$\varphi \wedge_{/W} \psi \equiv_U \psi \wedge_{/W} \varphi,$$

we can exploit the duality between $\models^+$ and $\models^-$ to save ourselves some effort. Normally, we would begin by showing that

$$\varphi \vee_{/W} \psi \equiv^+_U \psi \vee_{/W} \varphi \quad \text{and} \quad \varphi \vee_{/W} \psi \equiv^-_U \psi \vee_{/W} \varphi,$$

and then proceed to show

$$\varphi \wedge_{/W} \psi \equiv^+_U \psi \wedge_{/W} \varphi \quad \text{and} \quad \varphi \wedge_{/W} \psi \equiv^-_U \psi \wedge_{/W} \varphi.$$

However, the first truth equivalence holds for every pair φ and ψ of IF formulas if and only if the second falsity equivalence holds for every φ

and ψ. Likewise, the first falsity equivalence holds in general if and only if the second truth equivalence does too. Thus it suffices to prove that both truth equivalences hold for arbitrary φ and ψ.

5.3.2 Propositional laws

The failure of the law of excluded middle shows that not all of the propositional laws familiar to us from first-order logic are valid in IF logic. Remarkably, a restricted version of each propositional law does hold in IF logic with slashed connectives. We begin by verifying the law of double negation and De Morgan's laws.

Proposition 5.19 (Double negation) $\varphi \equiv_U \neg\neg\varphi$.

Proof   For any structure $\mathbb{M}$ and assignment team X with domain U we have

$$\mathbb{M}, X \models^{\pm} \neg\neg\varphi \quad \text{iff} \quad \mathbb{M}, X \models^{\mp} \neg\varphi$$
$$\text{iff} \quad \mathbb{M}, X \models^{\pm} \varphi. \qquad \dashv$$

Proposition 5.20 (De Morgan)

$$\neg(\varphi \vee_{/W} \psi) \equiv_U \neg\varphi \wedge_{/W} \neg\psi,$$
$$\neg(\varphi \wedge_{/W} \psi) \equiv_U \neg\varphi \vee_{/W} \neg\psi.$$

Proof   Recall that $\varphi \wedge_{/W} \psi$ is an abbreviation for $\neg(\neg\varphi \vee_{/W} \neg\psi)$. Hence

$$\neg(\varphi \vee_{/W} \psi) \equiv_U \neg(\neg\neg\varphi \vee_{/W} \neg\neg\psi) \equiv_U \neg\varphi \wedge_{/W} \neg\psi,$$

$$\neg(\varphi \wedge_{/W} \psi) \equiv_U \neg\neg(\neg\varphi \vee_{/W} \neg\psi) \equiv_U \neg\varphi \vee_{/W} \neg\psi. \qquad \dashv$$

Next we investigate the propositional laws corresponding to the standard axioms of Boolean algebra. Observe that commutativity is the only property that holds unconditionally.

Proposition 5.21   *Suppose $W \subseteq V$.*

Commutativity
$$\varphi \vee_{/W} \psi \equiv_U \psi \vee_{/W} \varphi,$$
$$\varphi \wedge_{/W} \psi \equiv_U \psi \wedge_{/W} \varphi.$$

Associativity
$$\varphi \vee_{/V} (\psi \vee_{/W} \chi) \models_U (\varphi \vee_{/V} \psi) \vee_{/W} \chi,$$
$$(\varphi \wedge_{/V} \psi) \wedge_{/W} \chi \models_U \varphi \wedge_{/V} (\psi \wedge_{/W} \chi).$$

Absorption

$$\varphi \vee_{/U} (\varphi \wedge_{/W} \psi) \equiv_U \varphi,$$
$$\varphi \wedge_{/U} (\varphi \vee_{/W} \psi) \equiv_U \varphi.$$

Distributivity

$$\varphi \vee_{/U} (\psi \wedge_{/U} \chi) \equiv_U (\varphi \vee_{/U} \psi) \wedge_{/U} (\varphi \vee_{/U} \chi),$$
$$\varphi \wedge_{/U} (\psi \vee_{/U} \chi) \equiv_U (\varphi \wedge_{/U} \psi) \vee_{/U} (\varphi \wedge_{/U} \chi).$$

Complementation

$$\varphi \vee_{/W} \neg\varphi \equiv_U^- \top,$$
$$\varphi \wedge_{/W} \neg\varphi \equiv_U^+ \bot.$$

Proof Let $\mathbb{M}$ be a suitable structure, and let X be a team of assignments with domain U.

(Commutativity) If $\mathbb{M}, X \models^+ \varphi \vee_{/W} \psi$ then

$$\mathbb{M}, Y \models^+ \varphi \quad \text{and} \quad \mathbb{M}, Y' \models^+ \psi$$

for some W-uniform cover $Y \cup Y' = X$, which implies $Y' \cup Y = X$ is a W-uniform cover such that

$$\mathbb{M}, Y' \models^+ \psi \quad \text{and} \quad \mathbb{M}, Y \models^+ \varphi.$$

Hence $\mathbb{M}, X \models^+ \psi \vee_{/W} \varphi$. The converse is symmetrical.

Observe that $\mathbb{M}, X \models^+ \varphi \wedge_{/W} \psi$ if and only if $\mathbb{M}, X \models^+ \varphi$ and $\mathbb{M}, X \models^+ \psi$ if and only if $\mathbb{M}, X \models^+ \psi \wedge_{/W} \varphi$.

(Associativity) Suppose $\mathbb{M}, X \models^+ \varphi \vee_{/V} (\psi \vee_{/W} \chi)$. Then for some V-uniform cover $Y \cup Y' = X$,

$$\mathbb{M}, Y \models^+ \varphi \quad \text{and} \quad \mathbb{M}, Y' \models^+ \psi \vee_{/W} \chi,$$

which implies there is a W-uniform cover $Y'' \cup Y''' = Y'$ such that $\mathbb{M}, Y'' \models^+ \psi$ and $\mathbb{M}, Y''' \models^+ \chi$. Without loss of generality we may assume Y, Y'', and Y''' are pairwise disjoint. To show Y and Y'' are V-saturated in their union, suppose $s \in Y$ and $s'' \in Y''$. Then $s'' \in Y'$, which implies $s \not\approx_V s''$. Thus $\mathbb{M}, Y \cup Y'' \models^+ \varphi \vee_{/V} \psi$. To show $Y \cup Y''$ and Y''' are W-saturated in X, suppose $s \in Y \cup Y''$ and $s''' \in Y'''$. If $s \in Y$, then we have $s \in Y$ and $s''' \in Y'''$, which implies $s \not\approx_V s'''$, so $s \not\approx_W s'''$, whereas if $s \in Y''$, then we have $s \in Y''$ and $s''' \in Y'''$, which implies $s \not\approx_W s'''$. Therefore $\mathbb{M}, X \models^+ (\varphi \vee_{/V} \psi) \vee_{/W} \chi$.

Observe that $\mathbb{M}, X \models^+ (\varphi \wedge_{/V} \psi) \wedge_{/W} \chi$ if and only if

$$\mathbb{M}, X \models^+ \varphi \quad \text{and} \quad \mathbb{M}, X \models^+ \psi \quad \text{and} \quad \mathbb{M}, X \models^+ \chi$$

if and only if $\mathbb{M}, X \models^+ \varphi \wedge_{/V} (\psi \wedge_{/W} \chi)$.

(Absorption) By Lemma 5.12(b), a cover $Y \cup Y' = X$ is uniform in U if and only if $Y = X$ or $Y' = X$. Thus $\mathbb{M}, X \models^+ \varphi \vee_{/U} (\varphi \wedge_{/W} \psi)$ implies $\mathbb{M}, X \models^+ \varphi$ or $\mathbb{M}, X \models^+ \varphi \wedge_{/W} \psi$. In either case $\mathbb{M}, X \models^+ \varphi$. Conversely, if $\mathbb{M}, X \models^+ \varphi$ then $X \cup \varnothing = X$ is a U-uniform cover such that $\mathbb{M}, X \models^+ \varphi$ and $\mathbb{M}, \varnothing \models^+ \varphi \wedge_{/W} \psi$. Hence

$$\mathbb{M}, X \models^+ \varphi \vee_{/U} (\varphi \wedge_{/W} \psi).$$

If $\mathbb{M}, X \models^+ \varphi \wedge_{/U} (\varphi \vee_{/W} \psi)$ then $\mathbb{M}, X \models^+ \varphi$ immediately. Conversely, if $\mathbb{M}, X \models^+ \varphi$, then $X \cup \varnothing = X$ is a W-uniform cover such that

$$\mathbb{M}, X \models^+ \varphi \quad \text{and} \quad \mathbb{M}, \varnothing \models^+ \psi.$$

Hence $\mathbb{M}, X \models^+ \varphi \vee_{/W} \psi$, which implies $\mathbb{M}, X \models^+ \varphi \wedge_{/U} (\varphi \vee_{/W} \psi)$.

(Distributivity) $\mathbb{M}, X \models^+ \varphi \vee_{/U} (\psi \wedge_{/U} \chi)$ if and only if $\mathbb{M}, X \models^+ \varphi$ or $\mathbb{M}, X \models^+ \psi \wedge_{/U} \chi$ if and only if $\mathbb{M}, X \models^+ \varphi \vee_{/U} \psi$ and $\mathbb{M}, X \models^+ \varphi \vee_{/U} \chi$ if and only if $\mathbb{M}, X \models^+ (\varphi \vee_{/U} \psi) \wedge_{/U} (\varphi \vee_{/U} \chi)$.

Similarly, $\mathbb{M}, X \models^+ \varphi \wedge_{/U} (\psi \vee_{/U} \chi)$ if and only if $\mathbb{M}, X \models^+ \varphi$ and $\mathbb{M}, X \models^+ \psi \vee_{/U} \chi$ if and only if $\mathbb{M}, X \models^+ \varphi \wedge_{/U} \psi$ or $\mathbb{M}, X \models^+ \varphi \wedge_{/U} \chi$ if and only if $\mathbb{M}, X \models^+ (\varphi \wedge_{/U} \psi) \vee_{/U} (\varphi \wedge_{/U} \chi)$.

(Complementation) $\mathbb{M}, X \models^+ \varphi \wedge_{/W} \neg \varphi$, if and only if $\mathbb{M}, X \models^+ \varphi$ and $\mathbb{M}, X \models^- \varphi$ if and only if $X = \varnothing$.   ⊣

The associative laws are one-directional. In the semantic game for

$$\varphi \vee_{/V} (\psi \vee_{/W} \chi)$$

Eloise is first prompted to choose between φ and $\psi \vee_{/W} \chi$, without seeing the variables in V. If she chooses the latter, she must then choose between ψ and χ, but only after the variables in $V - W$ are revealed to her. In the game for

$$(\varphi \vee_{/V} \psi) \vee_{/W} \chi$$

Eloise is prompted to choose between $\varphi \vee_{/V} \psi$ and χ without seeing the variables is W. If she chooses the former, she must then choose between φ and ψ, but only after the variables in $V - W$ are hidden. The associative law tells us that Eloise is better off making her initial choice with as much information as possible. When the slash sets are identical the associative law holds in both directions:

$$\varphi \vee_{/W} (\psi \vee_{/W} \chi) \equiv_U (\varphi \vee_{/W} \psi) \vee_{/W} \chi,$$
$$(\varphi \wedge_{/W} \psi) \wedge_{/W} \chi \equiv_U \varphi \wedge_{/W} (\psi \wedge_{/W} \chi).$$

5.3.3 Slash sets

Shrinking the slash set on a disjunction or existential quantifier helps
Eloise verify the formula by revealing the values of additional variables.
Shrinking the slash set on a conjunction or universal quantifier helps
Abelard falsify the formula.

Proposition 5.22 *Suppose $W \subseteq V$.*

(a) $\varphi \vee_{/V} \psi \models_U \varphi \vee_{/W} \psi$,
   $\varphi \wedge_{/W} \psi \models_U \varphi \wedge_{/V} \psi$.
(b) $(\exists x/V)\varphi \models_U (\exists x/W)\varphi$,
   $(\forall x/W)\varphi \models_U (\forall x/V)\varphi$.

Proof Let $\mathbb{M}$ be a suitable structure, and let X be a team of assignments
with domain U.

(a) Since $W \subseteq V$, every V-uniform cover $Y \cup Y' = X$ is automatically
uniform in W. Thus $\mathbb{M}, X \models^+ \varphi \vee_{/V} \psi$ implies $\mathbb{M}, X \models^+ \varphi \vee_{/W} \psi$.
Furthermore $\mathbb{M}, X \models^- \varphi \vee_{/V} \psi$ if and only if $\mathbb{M}, X \models^- \varphi \vee_{/W} \psi$ by
definition. Thus $\varphi \vee_{/V} \psi \models_U \varphi \vee_{/W} \psi$.

It follows from the previous statement that $\neg\varphi \vee_{/V} \neg\psi \models_U \neg\varphi \vee_{/W} \neg\psi$,
so by Proposition 5.16 we have $\varphi \wedge_{/W} \psi \models_U \varphi \wedge_{/V} \psi$.

(b) Similarly, since $W \subseteq V$, every V-uniform $f \colon X \to M$ is uniform in
W. Thus $\mathbb{M}, X \models^+ (\exists x/V)\varphi$ implies $\mathbb{M}, X \models^+ (\exists x/W)\varphi$. Furthermore,

$$\mathbb{M}, X \models^+ (\forall x/W)\varphi \quad \text{iff} \quad \mathbb{M}, X \models^+ (\forall x/V)\varphi$$

by definition. The dual follows from Proposition 5.16   ⊣

5.3.4 Distribution of quantifiers over connectives

In first-order logic existential quantifiers distribute over disjunctions —
and universal quantifiers distribute over conjunctions — because they
are both moves for the same player. In IF logic, quantifiers distribute
over similar connectives as long as the player has access to the same
information when making each choice.

Proposition 5.23 *Suppose $x \notin U$.*

$$(\exists x/W)(\varphi \vee_{/W} \psi) \equiv_U (\exists x/W)\varphi \vee_{/W} (\exists x/W)\psi,$$
$$(\forall x/W)(\varphi \wedge_{/W} \psi) \equiv_U (\forall x/W)\varphi \wedge_{/W} (\forall x/W)\psi.$$

Proof Let $\mathbb{M}$ be a suitable structure, and let X be a team of assignments with domain U. Suppose $\mathbb{M}, X \models^+ (\exists x/W)\varphi \vee_{/W} (\exists x/W)\psi$. Then there exists a W-uniform cover $X_1 \cup X_2 = X$ as well as two W-uniform functions $f_1 \colon X_1 \to M$ and $f_2 \colon X_2 \to M$ such that

$$\mathbb{M}, X_1[x, f_1] \models^+ \varphi \quad \text{and} \quad \mathbb{M}, X_2[x, f_2] \models^+ \psi.$$

Without loss of generality we may assume $X_1 \cap X_2 = \varnothing$, which allows us to define a W-uniform function $f \colon X \to M$ by

$$f(s) = \begin{cases} f_1(s) & \text{if } s \in X_1, \\ f_2(s) & \text{if } s \in X_2. \end{cases}$$

Since $x \notin U$, the cover $X_1[x, f_1] \cup X_2[x, f_2] = X[x, f]$ is uniform in W, which implies

$$\mathbb{M}, X \models^+ (\exists x/W)(\varphi \vee_{/W} \psi).$$

Conversely, suppose there exists a W-uniform function $f \colon X \to M$ and a W-uniform cover $Y_1 \cup Y_2 = X[x, f]$ such that $\mathbb{M}, Y_1 \models^+ \varphi$ and $\mathbb{M}, Y_2 \models^+ \psi$. Without loss of generality we may assume $Y_1 \cap Y_2 = \varnothing$. Define

$$X_1 = \{\, s \in X : s(x/f(s)) \in Y_1 \,\},$$
$$X_2 = \{\, s \in X : s(x/f(s)) \in Y_2 \,\},$$

and let f_i be the restriction of f to X_i. Then the fact that $Y_i = X_i[x, f_i]$ implies

$$\mathbb{M}, X_1 \models^+ (\exists x/W)\varphi \quad \text{and} \quad \mathbb{M}, X_2 \models^+ (\exists x/W)\psi.$$

Since the cover $X_1 \cup X_2 = X$ is uniform in W we have

$$\mathbb{M}, X \models^+ (\exists x/W)\varphi \vee_{/W} (\exists x/W)\psi.$$

Observe that $\mathbb{M}, X \models^+ (\forall x/W)(\varphi \wedge_{/W} \psi)$ if and only if

$$\mathbb{M}, X[x, M] \models^+ \varphi \quad \text{and} \quad \mathbb{M}, X[x, M] \models^+ \psi$$

if and only if $\mathbb{M}, X \models^+ (\forall x/W)\varphi \wedge_{/W} (\forall x/W)\psi$.   ⊣

When the quantifier and the connective belong to different players, the player who moves first is at a disadvantage.

Proposition 5.24 *Suppose $x \in W$ or $x \notin U$.*

$$(\exists x/V)(\varphi \wedge_{/W} \psi) \models_U (\exists x/V)\varphi \wedge_{/W} (\exists x/V)\psi,$$
$$(\forall x/V)\varphi \vee_{/W} (\forall x/V)\psi \models_U (\forall x/V)(\varphi \vee_{/W} \psi).$$

Proof Let $\mathbb{M}$ be a suitable structure, and let X be a team of assignments with domain U. If $\mathbb{M}, X \models^+ (\exists x/V)(\varphi \wedge_{/W} \psi)$, then there is a V-uniform function $f \colon X \to M$ such that $\mathbb{M}, X[x, f] \models^+ \varphi$ and $\mathbb{M}, X[x, f] \models^+ \psi$. Hence

$$\mathbb{M}, X \models^+ (\exists x/V)\varphi \wedge_{/W} (\exists x/V)\psi.$$

Instead of proving $(\exists x/V)\varphi \wedge_{/W} (\exists x/V)\psi \models^-_U (\exists x/V)(\varphi \wedge_{/W} \psi)$, it suffices to show

$$(\forall x/V)\varphi \vee_{/W} (\forall x/V)\psi \models^+_U (\forall x/V)(\varphi \vee_{/W} \psi).$$

If $\mathbb{M}, X \models^+ (\forall x/V)\varphi \vee_{/W} (\forall x/V)\psi$, then for some W-uniform cover $Y \cup Y' = X$ we have $\mathbb{M}, Y[x, M] \models^+ \varphi$ and $\mathbb{M}, Y'[x, M] \models^+ \psi$. The cover

$$Y[x, M] \cup Y'[x, M] = X[x, M]$$

is uniform in W because $x \in W$ or $x \notin U$. Thus

$$\mathbb{M}, X \models^+ (\forall x/V)(\varphi \vee_{/W} \psi). \qquad \dashv$$

The hypotheses $x \in W$ or $x \notin U$ are necessary because otherwise Eloise's choice between $(\forall x/V)\varphi$ and $(\forall x/V)\psi$ may depend on information encoded in x that gets erased by the universal quantifier in $(\forall x/V)(\varphi \vee_{/W} \psi)$.

For example, let $\mathbb{B} = \{0, 1\}$ be a two-element structure in which both elements are named by constant symbols, and let $s_{ij} = \{(x, i), (y, j)\}$. Observe that

$$\mathbb{B}, \{s_{00}, s_{11}\} \models^+ \forall x(y = 0) \vee_{/\{y\}} \forall x(y = 1)$$

because $\mathbb{B}, \{s_{00}\} \models^+ \forall x(y = 0)$ and $\mathbb{B}, \{s_{11}\} \models^+ \forall x(y = 1)$. However,

$$\mathbb{B}, \{s_{00}, s_{11}\} \not\models^+ \forall x(y = 0 \vee_{/\{y\}} y = 1)$$

because $\mathbb{B}, \{s_{00}, s_{01}, s_{10}, s_{11}\} \not\models^+ y = 0 \vee_{/\{y\}} y = 1$.

5.3.5 Vacuous quantifiers

Vacuous quantifiers do not affect the meaning of first-order formulas, but they can affect the meaning of IF formulas. If φ is a first-order formula, and x does not occur in φ, then $\varphi \equiv \exists x \varphi$ (see Proposition 3.13). In IF formulas, however, vacuous quantifiers create signaling opportunities — a player can assign a value to the new variable that signals the value of another (hidden) variable (see Examples 4.8 and 4.26). One way to prevent a vacuous quantifier from altering the meaning of an IF formula is to hide the value of the vacuously quantified variable.

Proposition 5.25 (Caicedo *et al.* [9, Theorem 11.1]) *If the variable x occurs neither in φ nor in U, then*

$$\varphi \equiv_U (Qx/W)\varphi/\{x\}.$$

Proof Let M be a suitable structure, let X be a team of assignments with domain U, and let $f\colon X \to M$ be any W-uniform function (e.g., a constant function). Then $X[x, f]$ is an extension of X to $\mathrm{dom}(X) \cup \{x\}$, so by Lemma 5.8,

$$M, X \models^+ \varphi \quad \text{implies} \quad M, X[x, f] \models^+ \varphi/\{x\}$$
$$\text{implies} \quad M, X \models^+ (\exists x/W)\varphi/\{x\}.$$

Conversely, if there is a W-uniform $f\colon X \to M$ such that $M, X[x, f] \models^+ \varphi/\{x\}$, then $M, X \models^+ \varphi$. Similarly,

$$M, X \models^+ \varphi \quad \text{iff} \quad M, X[x, M] \models^+ \varphi/\{x\}$$
$$\text{iff} \quad M, X \models^+ (\forall x/W)\varphi/\{x\}. \qquad \dashv$$

Applying Proposition 5.25 to the formula from Example 4.26 yields

$$(\exists y/\{x\})x = y \ \equiv_{\{x\}} \ Qz(\exists y/\{x, z\})x = y.$$

Another way to prevent a vacuous quantifier from affecting the meaning of an IF formula is to make its slash set so large that the vacuously quantified variable cannot store any useful information.[4]

Proposition 5.26 *If x occurs neither in φ nor U, then $\varphi \equiv_U (Qx/U)\varphi$.*

Proof Let M be a suitable structure, and let X be a team of assignments with domain U. For all $a \in M$, let $s_a = \{(x, a)\}$, and let $f_a\colon X \to M$ be the constant function defined by $f_a(s) = a$, which is clearly uniform in U. Then

$$X[x, f_a] = X \times \{s_a\},$$
$$X[x, M] = X \times \{\, s_a : a \in M \,\},$$

are both Cartesian extensions of X to $\mathrm{dom}(X) \cup \{x\}$. Thus, by the

---

[4] The reader may wish to compare Proposition 5.26 to [9, Theorem 11.3].

Cartesian extension theorem,

$$\mathbb{M}, X \models^+ \varphi \quad \text{iff} \quad \mathbb{M}, X[x, f_a] \models^+ \varphi$$
$$\text{iff} \quad \mathbb{M}, X \models^+ (\exists x/U)\varphi,$$
$$\mathbb{M}, X \models^+ \varphi \quad \text{iff} \quad \mathbb{M}, X[x, M] \models^+ \varphi$$
$$\text{iff} \quad \mathbb{M}, X \models^+ (\forall x/U)\varphi. \qquad \dashv$$

Applying Proposition 5.26 to our example formula yields

$$(\exists y/\{x\})x = y \equiv_{\{x\}} (Qz/\{x\})(\exists y/\{x\})x = y.$$

5.3.6 Double quantification

In first-order logic, quantifying a variable twice in succession is equivalent to quantifying it once, with the latter quantifier determining the meaning of the formula. In IF logic, the situation is more complicated because we must account for the information available to the players. If the same player chooses the value of x twice in a row, the first choice is redundant when (a) the player has access to less information when making the first choice than when making the second choice, or (b) the player cannot see the value of x when making the second choice.

Proposition 5.27 *(a) If $x \notin U$ and $V \supseteq W$ then*

$$(Qx/V)(Qx/W)\varphi \equiv_U (Qx/W)\varphi.$$

(b) Likewise, if $x \in W$ then

$$(Qx/V)(Qx/W)\varphi \equiv_U (Qx/W)\varphi.$$

Proof (a) Suppose $\mathbb{M}, X \models^+ (\exists x/V)(\exists x/W)\varphi$. Then there exist a V-uniform function $f: X \to M$ and a W-uniform function $g: X[x, f] \to M$ such that $\mathbb{M}, X[x, f][x, g] \models^+ \varphi$. If $V \supseteq W$, we can define a W-uniform function $h: X \to M$ by

$$h(s) = g\big(s(x/f(s))\big).$$

Since $X[x, f][x, g] = X[x, h]$, it follows that $\mathbb{M}, X \models^+ (\exists x/W)\varphi.$

Conversely, suppose there is a W-uniform function $h: X \to M$ such that $\mathbb{M}, X[x, h] \models^+ \varphi$, and let $f: X \to M$ be any V-uniform function (e.g., a constant function). If $x \notin U$, define a W-uniform function $g\colon X[x, f] \to M$ by

$$g\big(s(x/f(s))\big) = h(s).$$

Then $X[x, f][x, g] = X[x, h]$, so we have $\mathbb{M}, X \models^+ (\exists x/V)(\exists x/W)\varphi$.
Furthermore,

$$\mathbb{M}, X \models^+ (\forall x/W)(\forall x/W)\varphi \quad \text{iff} \quad \mathbb{M}, X[x, M][x, M] \models^+ \varphi$$
$$\text{iff} \quad \mathbb{M}, X[x, M] \models^+ \varphi$$
$$\text{iff} \quad \mathbb{M}, X \models^+ (\forall x/W)\varphi.$$

The proof of (b) is similar. ⊣

For the sake of comparison, notice that applying Proposition 5.27(a) to the formula $(\exists y/\{x\})x = y$ yields

$$(\exists y/\{x\})x = y \equiv_{\{x\}} (\exists y/\{x\})(\exists y/\{x\})x = y,$$

while applying Proposition 5.27(b) to $(\exists y/\{x\})x = y$ yields

$$(\exists y/\{x, y\})x = y \equiv_{\{x, y\}} \exists y(\exists y/\{x, y\})x = y.$$

When different players choose the value of the same variable which has not already been assigned a value, and nothing else happens in between, then the first player's choice is irrelevant.

Proposition 5.28 *Suppose $x \notin U$.*

$$(\exists x/V)(\forall x/W)\varphi \equiv_U (\forall x/W)\varphi,$$
$$(\forall x/V)(\exists x/W)\varphi \equiv_U (\exists x/W)\varphi.$$

Proof If $\mathbb{M}, X \models^+ (\exists x/V)(\forall x/W)\varphi$ then there exists a V-uniform $f\colon V \to M$ such that $\mathbb{M}, X[x, f][x, M] \models^+ \varphi$. Since

$$X[x, f][x, M] = X[x, M]$$

it follows that $\mathbb{M}, X \models^+ (\forall x/W)\varphi$. Conversely, if $\mathbb{M}, X[x, M] \models^+ \varphi$ then for any constant function $f\colon X \to M$ we have $\mathbb{M}, X[x, f][x, M] \models^+ \varphi$; hence

$$\mathbb{M}, X \models^+ (\exists x/V)(\forall x/W)\varphi.$$

Finally, since $x \notin U$ the Cartesian extension theorem gives us

$$\mathbb{M}, X \models^+ (\forall x/V)(\exists x/W)\varphi \quad \text{iff} \quad \mathbb{M}, X[x, M] \models^+ (\exists x/W)\varphi$$
$$\text{iff} \quad \mathbb{M}, X \models^+ (\exists x/W)\varphi. \qquad ⊣$$

Applying Proposition 5.28 to our example formula $(\exists y/\{x\})x = y$ shows that it is equivalent (relative to $\{x\}$) to both

$$\forall y(\exists y/\{x\})x = y \quad \text{and} \quad (\forall y/\{x\})(\exists y/\{x\})x = y,$$

which demonstrates that the amount of information available to Abelard when he chooses the value of y is irrelevant. If Abelard is allowed to see the value of x, then conceivably he could try to signal its value to Eloise. Alas, he is unlikely to do so, which means Eloise cannot trust the value of y as a signal.

5.3.7 Interchange of quantifiers

Now we consider the case when two adjacent quantifiers quantify distinct variables. In first-order logic, like quantifiers commute because they are moves for the same player, whereas $\exists x \forall y \varphi$ entails $\forall y \exists x \varphi$ because in the latter Eloise's choice of x depends on Abelard's choice of y, and in the former it does not. With IF logic we have the power to specify the variables upon which a given quantifier depends irrespective of its position in the formula. In particular, we can reorder any two adjacent quantifiers by adjusting their slash sets. For example, if $\text{Free}(\varphi) = \{x, y\}$,

$$\forall x(\exists y/\{x\})\varphi \equiv \exists y(\forall x/\{y\})\varphi.$$

Even though the above IF sentences are equivalent, their semantic games differ in the order the players make their moves. An alternative view is that two mutually independent moves are made simultaneously. Van Benthem [2, pp. 199–200] has observed that the above equivalence can be seen as an instance of the game-algebraic identity

$$(G \times H)\,;K = (H \times G)\,;K.$$

When applied to IF logic, this general principle yields:

Proposition 5.29 (Caicedo *et al.* [9, Theorem 13.1]) *If x and y are distinct variables not in U, then*

$$(Qx/V)(Q'y/W \cup \{x\})\varphi \equiv_U (Q'y/W)(Qx/V \cup \{y\})\varphi.$$

Proof Let $\mathbb{M}$ be a suitable structure and X a team of assignments with domain U. Suppose $\mathbb{M}, X \models^+ (\exists x/V)(\exists y/W \cup \{x\})\varphi$. Then there exist a V-uniform function $f\colon X \to M$ and a $W \cup \{x\}$-uniform function $g\colon X[x, f] \to M$ such that $\mathbb{M}, X[x, f][y, g] \models^+ \varphi$. Define a W-uniform function $g' \colon X \to M$ by

$$g'(s) = g\big(s(x/f(s))\big),$$

and a $V \cup \{y\}$-uniform function $f' \colon X[y, g'] \to M$ by

$$f'\big(s(y/g'(s))\big) = f(s).$$

Then $X[x, f][y, g] = X[y, g'][x, f']$, so

$$\mathbb{M}, X \models^+ (\exists y/W)(\exists x/V \cup \{y\})\varphi.$$

The converse is symmetrical.

   Observe that

$$\mathbb{M}, X \models^+ (\forall x/V)(\forall y/W \cup \{x\})\varphi$$
$$\text{iff} \quad \mathbb{M}, X[x, M][y, M] \models^+ \varphi$$
$$\text{iff} \quad \mathbb{M}, X[y, M][x, M] \models^+ \varphi$$
$$\text{iff} \quad \mathbb{M}, X \models^+ (\forall y/W)(\forall x/V \cup \{y\})\varphi.$$

Now suppose $\mathbb{M}, X \models^+ (\exists x/V)(\forall y/W \cup \{x\})\varphi$. Then there is a V-uniform $f \colon X \to M$ such that

$$\mathbb{M}, X[x, f][y, M] \models^+ \varphi.$$

Define a $V \cup \{y\}$-uniform function $f' \colon X[y, M] \to M$ by

$$f'\big(s(y/a)\big) = f(s).$$

Then $\mathbb{M}, X[y, M][x, f'] \models^+ \varphi$, hence

$$\mathbb{M}, X \models^+ (\forall y/W)(\exists x/V \cup \{y\})\varphi.$$

Conversely, if there exists a $V \cup \{y\}$-uniform function $f' \colon X[y, M] \to M$ such that $\mathbb{M}, X[y, M][x, f'] \models^+ \varphi$ we can define a V-uniform function $f \colon X \to M$ by

$$f(s) = f'\big(s(y/a)\big).$$

Thus $\mathbb{M}, X[x, f][y, M] \models^+ \varphi$, which implies

$$\mathbb{M}, X \models^+ (\exists x/V)(\forall y/W \cup \{x\})\varphi.$$

The proof of the dual is similar.   ⊣

5.3.8 Quantifier extraction

Recall that in first-order logic we may add a formula ψ to the scope of a quantifier as long as the quantified variable does not occur free in ψ:

$$Qx\varphi \circ \psi \equiv Qx(\varphi \circ \psi).$$

This allows us to pull all of the quantifiers to the front of a first-order formula, placing it in prenex normal form (see Proposition 3.21 and Theorem 3.26).

In IF logic, quantifier extraction is a more delicate matter because the extra variable x may affect the players' strategies for the semantic game of ψ. On the one hand, a player may be able to use x to signal the value of a hidden variable, thus gaining an advantage.

Example 5.30 If $\mathbb{M} = \{a, b\}$ is a two-element structure, and

$$s'_a = \{(z, a)\}, \qquad s'_{aa} = \{(x, a), (z, a)\},$$
$$s'_b = \{(z, b)\}, \qquad s'_{bb} = \{(x, b), (z, b)\},$$

then $\mathbb{M}, \{s'_a, s'_b\} \not\models^+ \exists x(x \neq x) \vee (\exists y/\{z\})\, y = z$ because the left disjunct is tautologically false, and the right disjunct is neither satisfied nor dissatisfied by $\{s'_a, s'_b\}$. However,

$$\mathbb{M}, \{s'_a, s'_b\} \models^+ \exists x \Big[x \neq x \vee (\exists y/\{z\})\, y = z \Big]$$

because $\mathbb{M}, \{s'_{aa}, s'_{bb}\} \models^+ (\exists y/\{z\})\, y = z$. ⊣

On the other hand, if x already has a value, assigning it a new value might erase information that was previously stored there.

Example 5.31 (Caicedo *et al.* [9, Example 7.2]) Let $\mathbb{M}$ and the various assignments be as in Example 5.30. Then

$$\mathbb{M}, \{s'_{aa}, s'_{bb}\} \models^+ \forall x(x \neq x) \vee (\exists y/\{z\})\, y = z,$$

but

$$\mathbb{M}, \{s'_{aa}, s'_{bb}\} \not\models^+ \forall x \Big[x \neq x \vee (\exists y/\{z\})\, y = z \Big],$$

because the universal quantifier erases the information about z that was stored in x. ⊣

We solve the first problem by adding x to all of the slash sets in ψ, thereby preventing it from affecting the players' strategies for ψ. We solve the second problem by excluding x from the domains of the teams used to evaluate the formulas (see Theorem 5.35).

We will prove the quantifier extraction laws for IF logic using Skolem semantics. In order to do so, we need to revisit the Skolemization procedure for first-order formulas. If φ and φ' are first-order formulas in different vocabularies, then it is unlikely they that are equivalent since φ may have models that are unsuitable for φ', and vice versa. However, if φ is

$$\exists y \big[R(x,y) \vee f(x,y) = z \big]$$

and φ' is $\exists y \big[R'(x,y) \vee f'(x,y) = z \big]$ there is a sense in which the two formulas *are* equivalent, since one formula can be obtained from the other simply be renaming relation and function symbols.

Definition 5.32 Let L and L' be first-order vocabularies. A *renaming* of L to L' is a bijection $\rho \colon L \to L'$ such that every relation symbol in L is sent to a relation symbol in L' with the same arity, and every function symbol in L is sent to a function symbol in L' with the same arity.

If φ is an IF_L formula, then $\rho(\varphi)$ is obtained by systematically replacing the relation and function symbols in φ with their images under ρ. If $\mathbb{M}$ is an L-structure, let $\rho(\mathbb{M})$ denote the L'-structure with the same universe whose interpretations of the relation symbols $\rho(R)$ and function symbols $\rho(f)$ are given by

$$\rho(R)^{\rho(\mathbb{M})} = R^{\mathbb{M}} \quad \text{and} \quad \rho(f)^{\rho(\mathbb{M})} = f^{\mathbb{M}}. \qquad \dashv$$

It is a straightforward exercise to show that for any assignment s,

$$\mathbb{M}, s \models \varphi \quad \text{iff} \quad \rho(\mathbb{M}), s \models \rho(\varphi).$$

Lemma 5.33 *Let φ be an IF_L formula. If the variables in V occur neither in φ nor in U, then the function*

$$\rho \colon L \cup \big\{ f_\psi : \psi \in \mathrm{Subf}_\exists(\varphi) \big\} \to L \cup \big\{ f_\psi : \psi \in \mathrm{Subf}_\exists(\varphi/V) \big\}$$

that fixes L and sends $f_{(\exists x/W)\chi}$ to $f_{(\exists x/W \cup V)\chi/V}$ is a renaming such that $\rho\big(\mathrm{Sk}_U(\varphi)\big) = \mathrm{Sk}_{U \cup V}(\varphi/V)$.

Proof The function ρ is a bijection because distinct existential subformulas of φ correspond to distinct existential subformula of φ/V, and every existential subformula of φ/V has the form $(\exists x/W \cup V)\chi/V$ for some existential subformula $(\exists x/W)\chi$ of φ. If φ is a literal, then both $\mathrm{Sk}_U(\varphi)$ and $\mathrm{Sk}_{U \cup V}(\varphi/V)$ are simply φ, and $\rho(\varphi) = \varphi$ because ρ fixes L.

Suppose φ is $\psi \circ \psi'$. Then by inductive hypothesis

$$\rho(\text{Sk}_U(\psi \circ \psi')) = \rho(\text{Sk}_U(\psi)) \circ \rho(\text{Sk}_U(\psi'))$$
$$= \text{Sk}_{U \cup V}(\psi/V) \circ \text{Sk}_{U \cup V}(\psi'/V)$$
$$= \text{Sk}_{U \cup V}((\psi \circ \psi')/V).$$

Suppose φ is $(\exists x/W)\psi$. Then by inductive hypothesis

$$\rho\Big(\text{Sk}_U\big((\exists x/W)\psi\big)\Big)$$
$$= \rho\Big(\text{Subst}\big(\text{Sk}_{U \cup \{x\}}(\psi), x, f_{(\exists x/W)\psi}(y_1, \ldots, y_n)\big)\Big)$$
$$= \text{Subst}\Big(\rho\big(\text{Sk}_{U \cup \{x\}}(\psi)\big), x, \rho\big(f_{(\exists x/W)\psi}(y_1, \ldots, y_n)\big)\Big)$$
$$= \text{Subst}\big(\text{Sk}_{U \cup V \cup \{x\}}(\psi), x, f_{(\exists x/W \cup V)\psi/V}(y_1, \ldots, y_n)\big)$$
$$= \text{Sk}_{U \cup V}\big((\exists x/W \cup V)\psi/V\big),$$

where $y_1, \ldots, y_n$ enumerates $U - W = (U \cup V) - (W \cup V)$.

Suppose φ is $(\forall x/W)\psi$. Then by inductive hypothesis

$$\rho\Big(\text{Sk}_U\big((\forall x/W)\psi\big)\Big) = \rho\big(\forall x\, \text{Sk}_{U \cup \{x\}}(\psi)\big)$$
$$= \forall x \rho\big(\text{Sk}_{U \cup \{x\}}(\psi)\big)$$
$$= \forall x\, \text{Sk}_{U \cup V \cup \{x\}}(\psi/V)$$
$$= \text{Sk}_{U \cup V}\big((\forall x/W \cup V)\psi/V\big). \qquad \dashv$$

Lemma 5.34 *Let φ and ψ be* IF$_L$ *formulas. If x occurs neither in ψ nor in U, then the function*

$$\rho\colon L \cup \Big\{ f_\chi : \chi \in \text{Subf}_\exists\big((Qx/W)\varphi \circ \psi\big) \Big\}$$
$$\to L \cup \Big\{ f_\chi : \chi \in \text{Subf}_\exists\big((Qx/W)[\varphi \circ \psi/\{x\}]\big) \Big\}$$

that fixes L and $\{ f_\chi : \chi \in \text{Subf}_\exists(\varphi) \}$, sends f_χ to $f_{\chi/\{x\}}$ for all $\chi \in \text{Subf}_\exists(\psi)$, and sends $f_{(\exists x/W)\varphi}$ to $f_{(\exists x/W)[\varphi \circ \psi/\{x\}]}$ (if $Q = \exists$) is a renaming such that

$$\rho\Big(\text{Sk}_U\big((Qx/W)\varphi \circ \psi\big)\Big) = \text{Sk}_U\Big((Qx/W)[\varphi \circ \psi/\{x\}]\Big).$$

Proof First we consider the existential case. By Lemma 5.33,

$$\rho\Big(\mathrm{Sk}_U\big((\exists x/W)\varphi \circ \psi\big)\Big)$$

$$= \rho\Big(\mathrm{Subst}\big(\mathrm{Sk}_{U\cup\{x\}}(\varphi), x, f_{(\exists x/W)\varphi}(y_1,\dots,y_n)\big) \circ \mathrm{Sk}_U(\psi)\Big)$$

$$= \mathrm{Subst}\big(\mathrm{Sk}_{U\cup\{x\}}(\varphi), x, f_{(\exists x/W)[\varphi \circ \psi/\{x\}]}(y_1,\dots,y_n)\big)$$
$$\quad \circ \mathrm{Sk}_{U\cup\{x\}}(\psi/\{x\})$$

$$= \mathrm{Subst}\Big(\mathrm{Sk}_{U\cup\{x\}}(\varphi \circ \psi/\{x\}), x, f_{(\exists x/W)[\varphi \circ \psi/\{x\}]}(y_1,\dots,y_n)\Big)$$

$$= \mathrm{Sk}_U\Big((\exists x/W)[\varphi \circ \psi/\{x\}]\Big).$$

Next we consider the universal case. By Lemma 5.33,

$$\rho\Big(\mathrm{Sk}_U\big((\forall x/W)\varphi \circ \psi\big)\Big) = \rho\big(\forall x\, \mathrm{Sk}_{U\cup\{x\}}(\varphi) \circ \mathrm{Sk}_U(\psi)\big)$$

$$= \forall x\, \mathrm{Sk}_{U\cup\{x\}}(\varphi) \circ \mathrm{Sk}_{U\cup\{x\}}(\psi/\{x\})$$

$$= \forall x\Big(\mathrm{Sk}_{U\cup\{x\}}(\varphi \circ \psi/\{x\})\Big)$$

$$= \mathrm{Sk}_U\Big((\forall x/W)[\varphi \circ \psi/\{x\}]\Big). \qquad \dashv$$

Now we can prove the quantifier extraction laws, which we adapted from Theorems 7.5 and 8.3 in [9].

Theorem 5.35 *Let φ and ψ be IF_L formulas. If x occurs neither in ψ nor in U, then*

$$(Qx/W)\varphi \circ \psi \equiv_U (Qx/W)[\varphi \circ \psi/\{x\}].$$

Proof Let $\mathbb{M}$ be a suitable structure, let X be a team of assignments with domain U, and let ρ be the renaming defined in the previous lemma. Suppose $\mathbb{M}, X \models^+ (Qx/W)\varphi \circ \psi$. Then there is an expansion $\mathbb{M}^*$ of $\mathbb{M}$ to the vocabulary

$$L \cup \Big\{ f_\chi : \chi \in \mathrm{Subf}_\exists\big((Qx/W)\varphi \circ \psi\big) \Big\}$$

such that for all $s \in X$ we have $\mathbb{M}^*, s \models \mathrm{Sk}_U\big((Qx/W)\varphi \circ \psi\big)$. Thus by the previous lemma $\rho(\mathbb{M}^*)$ is an expansion of $\mathbb{M}$ to the vocabulary

$$L \cup \Big\{ f_\chi : \chi \in \mathrm{Subf}_\exists\big((Qx/W)[\varphi \circ \psi/\{x\}]\big) \Big\}$$

such that for all $s \in X$,

$$\rho(\mathbb{M}^*), s \models \mathrm{Sk}_U\Big((Qx/W)[\varphi \circ \psi/\{x\}]\Big).$$

Hence $\mathbb{M}, X \models^+ (Qx/W)[\varphi \circ \psi/\{x\}]$. The converse is similar. $\qquad \dashv$

We can also extract quantifiers from IF formulas with slashed connectives.

Corollary 5.36 *If x occurs neither in ψ nor in U, then*

$$(Qx/V)\varphi \circ_{/W} \psi \equiv_U (Qx/V)\big[\varphi \circ_{/W\cup\{x\}} \psi/\{x\}\big].$$

Proof The formula $(Qx/V)\varphi \vee_{/W} \psi$ is an abbreviation for

$$(\exists z/W)\Big[\big(z = 0 \wedge (Qx/V \cup \{z\})\varphi/\{z\}\big) \vee \big(z = 1 \wedge \psi/\{z\}\big)\Big]$$

which by a double application of Theorem 5.35 is equivalent relative to U to

$$(\exists z/W)(Qx/V \cup \{z\})\Big[(z = 0 \wedge \varphi/\{z\}) \vee (z = 1 \wedge \psi/\{x, z\})\Big].$$

By Proposition 5.29, the previous formula is equivalent relative to U to

$$(Qx/V)(\exists z/W \cup \{x\})\Big[(z = 0 \wedge \varphi/\{z\}) \vee (z = 1 \wedge \psi/\{x, z\})\Big]$$

which is $(Qx/V)\big[\varphi \vee_{/W\cup\{x\}} \psi/\{x\}\big]$.
Furthermore,

$$
\begin{aligned}
(Qx/V)\varphi \wedge_{/W} \psi &\equiv_U \neg\big[\neg(Qx/V)\varphi \vee_{/W} \neg\psi\big]\\
&\equiv_U \neg\big[(\overline{Q}x/V)\neg\varphi \vee_{/W} \neg\psi\big]\\
&\equiv_U \neg(\overline{Q}x/V)\big[\neg\varphi \vee_{/W\cup\{x\}} \neg\psi/\{x\}\big]\\
&\equiv_U \neg(\overline{Q}x/V)\neg\big[\varphi \wedge_{/W\cup\{x\}} \psi/\{x\}\big]\\
&\equiv_U (Qx/V)\big[\varphi \wedge_{/W\cup\{x\}} \psi/\{x\}\big]. \qquad \dashv
\end{aligned}
$$

5.3.9 Prenex normal form

An IF formula is in if all its quantifiers occur at the front of the formula, i.e., it has the form

$$(Q_1y_1/W_1)\ldots(Q_ny_n/W_n)\varphi,$$

where φ is quantifier free. To put an arbitrary IF formula in prenex normal form we must use the quantifier extraction laws to pull each quantifier to the front of the formula. Before we start applying Theorem 5.35 we must ensure that its hypotheses will be satisfied at each step. We can achieve this by renaming the bound variables of the formula so that every variable is quantified at most once, and no variable occurs both free and bound.

Renaming bound variables in IF formulas is a more delicate matter than in first-order formulas because we must take care not to alter the dependencies between the variables. For example, consider an IF sentence of the form

$$(\exists x/V)[\varphi \wedge (\exists y/W)\psi].$$

If the variable y occurs in φ, then we cannot use Theorem 5.35 to extract the quantifier $(\exists y/W)$. First, we need to rename the variable in $(\exists y/W)$ using a fresh variable z and replace every occurrence of y in the scope of $(\exists y/W)$ with an occurrence of z. If y happens to occur bound in ψ, then the bound occurrences of y will not be replaced when we substitute z for y using the usual substitution operation $\mathrm{Subst}(\psi, y, z)$. To avoid this problem, we work from inside out, starting with the innermost quantifier of a variable we wish to rename. Let us assume then that y does not occur bound in ψ. After renaming we obtain the formula

$$(\exists x/V)[\varphi \wedge (\exists z/W)\,\mathrm{Subst}(\psi, y, z)].$$

Unfortunately, this formula may not be equivalent to the previous one because if y is free in φ, then y will already have a value when play reaches the subformula $(\exists y/W)\psi$ or $(\exists z/W)\,\mathrm{Subst}(\psi, y, z)$. In the first case the value of y is overwritten when Eloise chooses a y independent of W, whereas in the second case it is not. To compensate for this assymmetry we add y to the slash sets in $\mathrm{Subst}(\psi, y, z)$:

$$(\exists x/V)\Big[\varphi \wedge (\exists z/W)\big(\mathrm{Subst}(\psi, y, z)/\{y\}\big)\Big].$$

The next proposition is adapted from [9, Theorems 6.12 and 6.13].

Proposition 5.37 *Suppose x does not occur bound in φ. If y occurs neither in $(Qx/W)\varphi$ nor in U, then*

$$(Qx/W)\varphi \equiv_U \begin{cases} (Qy/W)\,\mathrm{Subst}(\varphi, x, y) & \text{if } x \notin U, \\ (Qy/W)\big[\mathrm{Subst}(\varphi, x, y)/\{x\}\big] & \text{if } x \in U. \end{cases}$$

Proof Let $\mathbb{M}$ be a suitable structure, and let X be a team of assignments with domain U. By definition $\mathbb{M}, X \models^+ (\exists x/W)\varphi$ if and only if there exists a W-uniform function $f\colon X \to M$ such that $\mathbb{M}, X[x, f] \models^+ \varphi$, which holds by Lemma 5.4 if and only if

$$\mathbb{M}, \Big\{ s\big(y/f(s)\big)_{-x} : s \in X \Big\} \models^+ \mathrm{Subst}(\varphi, x, y). \tag{5.1}$$

If $x \notin U$ the above assignment team is equal to $X[y, f]$. Thus (5.1) is equivalent to $\mathbb{M}, X \models^+ (\exists y/W)\,\mathrm{Subst}(\varphi, x, y)$.

If $x \in U$ then $X[y, f]$ is an extension of the assignment team in (5.1) to $\mathrm{dom}(X) \cup \{y\}$. Consequently (5.1) holds if and only if

$$\mathbb{M}, X[y, f] \models^{+} \mathrm{Subst}(\varphi, x, y)/\{x\},$$

by Lemma 5.8. Thus (5.1) is equivalent to

$$\mathbb{M}, X \models^{+} (\exists y/W)\, \mathrm{Subst}(\varphi, x, y)/\{x\}.$$

For the universal case, first suppose $x \notin U$. Then by Lemma 5.4,

$$
\begin{aligned}
\mathbb{M}, X \models^{+} (\forall x/W)\varphi \quad &\text{iff} \quad \mathbb{M}, X[x, M] \models^{+} \varphi \\
&\text{iff} \quad \mathbb{M}, X[y, M] \models^{+} \mathrm{Subst}(\varphi, x, y) \\
&\text{iff} \quad \mathbb{M}, X \models^{+} (\forall y/W)\, \mathrm{Subst}(\varphi, x, y),
\end{aligned}
$$

Now suppose $x \in U$. Then by Lemmas 5.4 and 5.8,

$$
\begin{aligned}
\mathbb{M}, X \models^{+} (\forall x/W)\varphi & \\
\text{iff} \quad & \mathbb{M}, X[x, M] \models^{+} \varphi \\
\text{iff} \quad & \mathbb{M}, \{\, s(y/a)_{-x} : s \in X,\, a \in M \,\} \models^{+} \mathrm{Subst}(\varphi, x, y) \\
\text{iff} \quad & \mathbb{M}, X[y, M] \models^{+} \mathrm{Subst}(\varphi, x, y)/\{x\} \\
\text{iff} \quad & \mathbb{M}, X \models^{+} (\forall y/W)\big[\mathrm{Subst}(\varphi, x, y)/\{x\}\big]. \qquad \dashv
\end{aligned}
$$

To show the hypothesis $y \notin U$ is necessary, consider the following example adapted from [9, p. 111]. Let φ be the formula

$$\exists x_2 (\exists x_3/\{x_1\})(x_0 = x_2 \wedge x_1 = x_3),$$

and let φ' be the formula

$$\exists x_4 (\exists x_3/\{x_1\})(x_0 = x_4 \wedge x_1 = x_3).$$

Let ψ and ψ' denote the subformulas of φ and φ', respectively, obtained by removing each formula's initial quantifier. Notice that φ' is $\exists x_4\, \mathrm{Subst}(\psi, x_2, x_4)$. Now let $\mathbb{M} = \{a, b\}$ be a two-element structure, and

$$
\begin{aligned}
s_{aaa} &= \big\{(x_0, a), (x_1, a), (x_4, a)\big\}, \\
s_{abb} &= \big\{(x_0, a), (x_1, b), (x_4, b)\big\}, \\
s_{baa} &= \big\{(x_0, b), (x_1, a), (x_4, a)\big\}, \\
s_{bbb} &= \big\{(x_0, b), (x_1, b), (x_4, b)\big\}.
\end{aligned}
$$

One can verify that

$$\mathbb{M}, \{s_{aaa}, s_{abb}, s_{baa}, s_{bbb}\} \models^+ \varphi,$$
$$\mathbb{M}, \{s_{aaa}, s_{abb}, s_{baa}, s_{bbb}\} \not\models^+ \varphi'.$$

We now return to our main example. After renaming y to a fresh variable z we can extract $(\exists z/W)$, yielding either

$$(\exists x/V)(\exists z/W)\big[\varphi \wedge \mathrm{Subst}(\psi, y, z)\big] \quad \text{or}$$
$$(\exists x/V)(\exists z/W)\big[\varphi \wedge \mathrm{Subst}(\psi, y, z)/\{y\}\big]$$

depending on whether y belongs to the domain of the assignment team used to evaluate the formula.

Theorem 5.38 (Caicedo *et al.* [9, Theorems 9.3 and 9.4]) *Let φ be an IF_L formula with n quantifiers. For every finite set of variables U that contains $\mathrm{Free}(\varphi)$, and every set of variables V of size n that is disjoint from U, there exists an IF_L formula $\varphi' \equiv_U \varphi$ such that $\mathrm{Free}(\varphi) \subseteq \mathrm{Free}(\varphi') \subseteq U$ and $\mathrm{Bound}(\varphi') = V$.*

Proof Let U be a finite set of variables containing $\mathrm{Free}(\varphi)$, and let V be a set of n variables such that $U \cap V = \varnothing$. If φ is a literal there is nothing to prove.

Suppose φ is $\psi_1 \circ \psi_2$, where ψ_1 has m_1 quantifiers, and ψ_2 has m_2 quantifiers. Let $V_1 \cup V_2 = V$ be a disjoint cover of V such that $|V_1| = m_1$ and $|V_2| = m_2$. By inductive hypothesis, there exist IF_L formulas $\psi_1' \equiv_U \psi_1$ and $\psi_2' \equiv_U \psi_2$ such that $\mathrm{Free}(\psi_i) \subseteq \mathrm{Free}(\psi_i') \subseteq U$ and $\mathrm{Bound}(\psi_i') = V_i$. It follows that $\psi_1' \circ \psi_2' \equiv_U \psi_1 \circ \psi_2$,

$$\mathrm{Free}(\psi_1 \circ \psi_2) \subseteq \mathrm{Free}(\psi_1' \circ \psi_2') \subseteq U,$$

and $\mathrm{Bound}(\psi_1' \circ \psi_2') = V$.

Suppose φ is $(Qx/W)\psi$, where $x \notin U$. Then $\mathrm{Free}(\psi) \subseteq U \cup \{x\}$, so for any subset $V' \subseteq V$ of size $n-1$ that does not include x, there exists an IF_L formula $\psi' \equiv_{U \cup \{x\}} \psi$ such that

$$\mathrm{Free}(\psi) \subseteq \mathrm{Free}(\psi') \subseteq U \cup \{x\}$$

and $\mathrm{Bound}(\psi') = V'$. If $V = V' \cup \{x\}$, then $(Qx/W)\psi' \equiv_U (Qx/W)\psi$,

$$\mathrm{Free}\big((Qx/W)\psi\big) \subseteq \mathrm{Free}\big((Qx/W)\psi'\big) \subseteq U,$$

and $\mathrm{Bound}\big((Qx/W)\psi'\big) = V$.

If $V = V' \cup \{y\}$, where y is a variable distinct from x, let ψ'' be Subst(ψ', x, y). Then by Proposition 5.37,

$$(Qx/W)\psi \equiv_U (Qx/W)\psi' \equiv_U (Qy/W)\psi''.$$

Furthermore, since $y \notin \text{Free}(\psi)$ and $\text{Free}(\psi) - \{x\} \subseteq \text{Free}(\psi'') \subseteq U \cup \{y\}$ we have

$$\text{Free}((Qx/W)\psi) \subseteq \text{Free}((Qy/W)\psi'') \subseteq U.$$

Also, $\text{Bound}((Qy/W)\psi'') = V$ because $\text{Bound}(\psi'') = \text{Bound}(\psi') = V'$.

Now suppose φ is $(Qx/W)\psi$, where $x \in U$. Then $\text{Free}(\psi) \subseteq U$, so for any subset of $V' \subseteq V$ of size $n-1$, there is an IF$_L$ formula $\psi' \equiv_U \psi$ such that $\text{Free}(\psi) \subseteq \text{Free}(\psi') \subseteq U$ and $\text{Bound}(\psi') = V'$. Let $V - V' = \{y\}$, and let ψ'' be Subst$(\psi', x, y)/\{x\}$. Then by Proposition 5.37,

$$(Qx/W)\psi \equiv_U (Qx/W)\psi' \equiv_U (Qy/W)\psi''.$$

Finally, by the same argument as above,

$$\text{Free}((Qx/W)\psi) \subseteq \text{Free}((Qy/W)\psi'') \subseteq U$$

and $\text{Bound}((Qy/W)\psi'') = V$. ⊣

Since the formula φ' has the same number of quantifiers as bound variables, every bound variable must be quantified exactly once. Thus, the previous theorem implies that every IF formula φ is equivalent (relative to U) to an IF formula φ' in which every variable is quantified at most once, and no variable occurs both free and bound. If $U = \text{Free}(\varphi)$, then φ' will have the same free variables as φ.

If $\text{Free}(\varphi) \subset U$, then φ' may not have the same free variables as φ. Consider the IF formula

$$P(y) \wedge \forall x \exists y R(x, y).$$

If $U = \{x, y\}$ and $V = \{u, v\}$, when we use Proposition 5.37 to rename x to u and y to v we obtain

$$P(y) \wedge \forall u (\exists v / \{x\}) R(u, v).$$

Notice that x is free in the above formula, but not in the original formula.

As an exercise in applying Theorem 5.38, the reader should show that $\forall x \forall x \forall x P(x)$ is equivalent to $\forall u \forall v (\forall w / \{u\}) P(w)$.

Theorem 5.39 *Let φ be an IF$_L$ formula in which every variable is quantified at most once, and no variable occurs both free and bound. For every finite set of variables U that contains $\text{Free}(\varphi)$ and is disjoint from*

Bound(φ), *there exists an* IF_L *formula* $\varphi^* \equiv_U \varphi$ *in prenex normal form with the same free and bound variables as* φ.

Proof If φ is quantifier free, then it is in prenex normal form. Otherwise, let $(\exists x/W)$ be the left-most quantifier in φ. We use Theorem 5.35 to pull $(\exists x/W)$ to the front of the formula. To illustrate the procedure, let us assume φ has the form

$$((Qx/W)\psi \circ \psi') \circ \psi''.$$

By hypothesis, the variable x does not occur in ψ', ψ'', or U. Thus, applying Theorem 5.35 once yields

$$\Big((Qx/W)[\psi \circ \psi'/\{x\}]\Big) \circ \psi''.$$

Applying Theorem 5.35 again yields

$$(Qx/W)\Big[(\psi \circ \psi'/\{x\}) \circ \psi''/\{x\}\Big].$$

Observe that the formula inside the square brackets has fewer quantifiers than φ, its free variables are contained in $U \cup \{x\}$, and x does not occur bound, so by inductive hypothesis there is an equivalent (relative to $U \cup \{x\}$) formula ψ^* in prenex normal form with the same free and bound variables. Thus $\varphi \equiv_U (\exists x/W)\psi^*$, which is in prenex normal form and has the same free and bound variables as φ.   ⊣

Corollary 5.40 (Caicedo *et al.* [9, Theorem 10.1]) *Let* φ *be an* IF_L *formula with* n *quantifiers. For every finite set of variables* U *that contains* $\mathrm{Free}(\varphi)$, *and every set of variables* V *of size* n *that is disjoint from* U, *there exists an* IF_L *formula* $\varphi^* \equiv \varphi$ *in prenex normal form such that* $\mathrm{Free}(\varphi) \subseteq \mathrm{Free}(\varphi^*) \subseteq U$ *and* $\mathrm{Bound}(\varphi^*) = V$.

Proof Let U be a finite set of variables containing $\mathrm{Free}(\varphi)$, and let V be a set of n variables such that $U \cap V = \varnothing$. By Theorem 5.38, there exists an IF_L formula $\varphi' \equiv_U \varphi$ such that $\mathrm{Free}(\varphi) \subseteq \mathrm{Free}(\varphi') \subseteq U$ and $\mathrm{Bound}(\varphi') = V$. By Theorem 5.39, there exists an IF_L formula $\varphi^* \equiv_U \varphi'$ in prenex normal form such that $\mathrm{Free}(\varphi) \subseteq \mathrm{Free}(\varphi^*) \subseteq U$ and $\mathrm{Bound}(\varphi^*) = V$.   ⊣

Corollary 5.41 (Caicedo *et al.* [9, Corollary 10.3]) *Every* IF_L *sentence is equivalent to an* IF_L *sentence in prenex normal form.*

Consider the IF sentence

$$\forall x \exists y \Big[P(y) \wedge \forall x \exists y R(x,y) \wedge (\forall z/\{y\})R(y,z)\Big].$$

In order to place the sentence in prenex normal form, we first rename bound variables:

$$\forall x \exists y \Big[P(y) \wedge \forall u (\exists v / \{x\}) R(u, v) \wedge (\forall z / \{y\}) R(y, z) \Big].$$

Then we extract the quantifiers, one after another:

$$\forall x \exists y \forall u \Big[P(y) \wedge (\exists v / \{x\}) R(u, v) \wedge (\forall z / \{y, u\}) R(y, z) \Big],$$

$$\forall x \exists y \forall u (\exists v / \{x\}) \Big[P(y) \wedge R(u, v) \wedge (\forall z / \{y, u, v\}) R(y, z) \Big],$$

$$\forall x \exists y \forall u (\exists v / \{x\}) (\forall z / \{y, u, v\}) \Big[P(y) \wedge R(u, v) \wedge R(y, z) \Big].$$

5.3.10 Hintikka normal form

Most of the time our sympathies lie with Eloise — we are often more interested in whether an IF sentence is true than whether it is false. In such cases, we can ignore the slash sets on universal quantifiers because limiting the information available to Abelard affects the strategies he may follow, but it does not affect which actions he is allowed to perform. Thus, Eloise must plan for the same eventualities, regardless of whether Abelard makes his moves with full knowledge or in complete ignorance. For example, consider the IF sentences

$$\exists x \forall y (\exists z / \{y\}) R(x, y, z) \quad \text{and} \quad \exists x (\forall y / \{x\}) (\exists z / \{y\}) R(x, y, z).$$

Both sentences have the same Skolemization, namely

$$\forall y R (c, y, f(c)).$$

Hence, Eloise has a winning strategy for one if and only if she has a winning strategy for the other.

Dually, we can modify the slash sets on existential quantifiers without affecting whether Abelard has a winning strategy.

Lemma 5.42 $(\forall x / V) \varphi \equiv_U^+ (\forall x / W) \varphi$ *and* $(\exists x / V) \varphi \equiv_U^- (\exists x / W) \varphi$.

Proof Let $\mathbb{M}$ be a suitable structure, and X a team of assignments with domain U. Then $\mathbb{M}, X \models^+ (\forall x / V) \varphi$ if and only if $\mathbb{M}, X[x, M] \models^+ \varphi$ if and only if $\mathbb{M}, X \models^+ (\forall x / W) \varphi$. $\dashv$

Lemma 5.42 allows us to strengthen Proposition 5.29 as follows:

Proposition 5.43   *If x and y are distinct variables not in U, then*

$$(\exists x/V)(\forall y/W)\varphi \equiv_U^+ \forall y(\exists x/V \cup \{y\})\varphi.$$

Proof   By Lemma 5.42 and Proposition 5.29,

$$
\begin{aligned}
(\exists x/V)(\forall y/W)\varphi &\equiv_U^+ (\exists x/V)(\forall y/W \cup \{x\})\varphi \\
&\equiv_U (\forall y/W)(\exists x/V \cup \{y\})\varphi \\
&\equiv_U^+ \forall y(\exists x/V \cup \{y\})\varphi.
\end{aligned}
$$
$\dashv$

In those cases when we only care about the satisfaction (as opposed to dissatisfaction) of a particular IF formula, we can use Proposition 5.43 to obtain an improved prenex normal form theorem.

Definition 5.44   An IF formula is in *Hintikka normal form* if it is in prenex normal form, every universal quantifier is superordinate to every existential quantifier, and all of its universal quantifiers are unslashed, i.e., it has the form

$$\forall y_1 \ldots \forall y_m(\exists y_{m+1}/W_{m+1})\ldots(\exists y_n/W_n)\varphi$$

where φ is quantifier free. $\dashv$

Theorem 5.45   *Let φ be an IF_L formula. For every finite set of variables U that contains $\mathrm{Free}(\varphi)$, there exists an IF_L formula $\varphi^{**} \equiv_U^+ \varphi$ in Hintikka normal form such that $\mathrm{Free}(\varphi) \subseteq \mathrm{Free}(\varphi^{**}) \subseteq U$.*

Proof   By Corollary 5.40, there is an IF_L formula $\varphi^* \equiv_U \varphi$ in prenex normal form such that $\mathrm{Free}(\varphi) \subseteq \mathrm{Free}(\varphi^*) \subseteq U$. To obtain the desired IF_L formula φ^{**} in Hintikka normal form, use Proposition 5.43 to pull every universal quantifier in front of every existential quantifier. $\dashv$

Corollary 5.46   *Every IF_L sentence is truth equivalent to an IF_L sentence in Hintikka normal form.*

For example, when we place the sentence

$$\forall x \exists y \forall u(\exists v/\{x\})(\forall z/\{y,u,v\})\Big[P(y) \wedge R(u,v) \wedge R(y,z)\Big].$$

in Hintikka normal form, we get

$$\forall x \forall u \forall z(\exists y/\{u,z\})(\exists v/\{x,z\})\Big[P(y) \wedge R(u,v) \wedge R(y,z)\Big].$$

5.4 Model theory

So far, we have only considered one formula or sentence at a time. In this section, we consider sets of IF sentences. Many familiar theorems from first-order logic such as the compactness theorem and the Löwenheim-Skolem theorem lift to IF logic via Skolem semantics.

Definition 5.47   A set of sentences in some logical language is called a *theory*. If Γ is a first- or second-order theory, a structure $\mathbb{M}$ *models* Γ if it models every sentence in Γ. In symbols, $\mathbb{M} \models \Gamma$ if and only if for every $\varphi \in \Gamma$ we have $\mathbb{M} \models \varphi$.   ⊣

Definition 5.48   If Γ is an IF theory, then Γ is *true* in $\mathbb{M}$, written $\mathbb{M} \models^+ \Gamma$, if for all $\varphi \in \Gamma$ we have $\mathbb{M} \models^+ \varphi$, and Γ is *false* in $\mathbb{M}$, written $\mathbb{M} \models^- \Gamma$, if for all $\varphi \in \Gamma$ we have $\mathbb{M} \models^- \varphi$.   ⊣

A first- or second-order theory Γ is *satisfiable* if it has a model in the above sense, i.e., if there is a structure $\mathbb{M}$ such that $\mathbb{M} \models \Gamma$. An IF theory Γ is *satisfiable* if there is a structure such that $\mathbb{M} \models^+ \Gamma$.

5.4.1 Compactness

The well-known compactness theorem for first-order logic states that a first-order theory Γ has a model if every finite subtheory of Γ has a model. A similar theorem holds for IF logic.

Theorem 5.49 (Compactness)   *An IF theory Γ is satisfiable if every finite subtheory of Γ is satisfiable.*

Proof   Observe that by the Skolem semantics for IF logic (Definition 4.12), an IF theory Γ is satisfiable if and only if

$$\Gamma^* = \big\{ \, \mathrm{Sk}(\varphi) : \varphi \in \Gamma \, \big\}$$

is satisfiable. Hence, if every finite subtheory $\Delta \subseteq \Gamma$ is satisfiable, then so is every finite subtheory $\Delta^* \subseteq \Gamma^*$. By the compactness theorem for first-order logic, Γ^* must be satisfiable, which implies Γ is satisfiable too.   ⊣

There is a stronger version of compactness that holds in first-order logic, but not IF logic.

Definition 5.50   When $\Gamma \cup \{\varphi\}$ is a first- or second-order theory, Γ *entails* φ, denoted $\Gamma \models \varphi$, if for every suitable structure $\mathbb{M}$,

$$\mathbb{M} \models \Gamma \quad \text{implies} \quad \mathbb{M} \models \varphi.$$

When Γ is empty we simply write $\models \varphi$.   ⊣

Definition 5.51   When $\Gamma \cup \{\varphi\}$ is an IF theory, Γ *truth entails* φ, denoted $\Gamma \models^+ \varphi$, if

$$\mathbb{M} \models^+ \Gamma \quad \text{implies} \quad \mathbb{M} \models^+ \varphi.$$

Γ *falsity entails* φ, denoted $\Gamma \models^- \varphi$, if

$$\mathbb{M} \models^- \Gamma \quad \text{implies} \quad \mathbb{M} \models^- \varphi.$$

When Γ is empty, we simply write $\models^+ \varphi$ or $\models^- \varphi$, as appropriate.   ⊣

An alternative formulation of the compactness theorem for first-order logic is the following: Every first-order theory $\Gamma \cup \{\varphi\}$ has the property that $\Gamma \models \varphi$ if and only if there exists a finite $\Delta \subseteq \Gamma$ such that $\Delta \models \varphi$. In contrast, when Γ is an IF theory it is possible to have $\Gamma \models^+ \varphi$ even if $\Delta \not\models^+ \varphi$ for every finite $\Delta \subseteq \Gamma$.

Example 5.52   Let φ_n denote the IF sentence

$$\exists x_1 \dots \exists x_n \left(\bigwedge_{1 \le i < j \le n} x_i \ne x_j \right)$$

which asserts that the universe has at least n elements. Then

$$\{\, \varphi_n : n \ge 2 \,\} \models^+ \varphi_\infty,$$

where φ_∞ is the IF sentence that asserts the universe is infinite (see Example 4.14). However, there is no finite subtheory $\Delta \subseteq \{\, \varphi_n : n \ge 2 \,\}$ such that $\Delta \models^+ \varphi_\infty$.   ⊣

It follows immediately from the previous example that IF cannot have a complete proof system in which proofs have finite length.

Theorem 5.53   *There is no sound and semantically complete proof system for IF logic. That is, there is no proof system* $\vdash_{\mathrm{IF}}$ *such that for every IF theory* $\Gamma \cup \{\varphi\}$,

$$\Gamma \vdash_{\mathrm{IF}} \varphi \quad \textit{iff} \quad \Gamma \models^+ \varphi.$$

Proof Suppose for the sake of a contradiction that $\vdash_{\text{IF}}$ is such a proof system. Then by Example 5.52 we must have

$$\{\, \varphi_n : n \geq 2 \,\} \vdash_{\text{IF}} \varphi_\infty.$$

Since a proof of φ_∞ from $\{\, \varphi_n : n \geq 2 \,\}$ can use at most finitely many premises, there must be a finite subtheory $\Delta \subseteq \{\, \varphi_n : n \geq 2 \,\}$ such that $\Delta \vdash_{\text{IF}} \varphi_\infty$, which would imply $\Delta \models^+ \varphi_\infty$.   ⊣

A proof system $\vdash$ for a logical language is *weakly complete* if for every sentence φ in the language we have $\vdash \varphi$ if and only if $\models \varphi$. In other words, every valid sentence is provable. One naturally wonders whether IF logic might have a proof system that is complete in this weaker sense.

Theorem 5.54 *There is no proof system $\vdash_{\text{IF}}$ such that for every IF sentence φ we have $\vdash_{\text{IF}} \varphi$ if and only if $\models^+ \varphi$.*

Proof Observe that for every IF sentence φ we have $\models^+ \varphi \vee \varphi_\infty$ if and only if φ is true in every (suitable) finite model. Thus, if there were such a proof system, the set of first-order sentences that are true in every (suitable) finite model would be recursively enumerable, contrary to Trakhtenbrot's theorem.[5]   ⊣

5.4.2 The Löwenheim-Skolem theorem

The Löwenheim-Skolem theorem states that if a countable first-order theory has an infinite model, then it has models of every infinite cardinality [55–57, 61]. Like the compactness theorem, we can extend the Löwenheim-Skolem theorem to IF logic.

Theorem 5.55 (Löwenheim-Skolem) *Let Γ be a countable IF theory. If there is an infinite structure $\mathbb{M}$ such that $\mathbb{M} \models^+ \Gamma$, then for all infinite cardinalities κ there is a structure $\mathbb{M}'$ of size κ such that $\mathbb{M}' \models^+ \Gamma$.*

Proof Suppose $\mathbb{M}$ is an infinite structure such that $\mathbb{M} \models^+ \Gamma$ and that κ is an infinite cardinal. Then there is an expansion $\mathbb{M}^*$ of $\mathbb{M}$ such that $\mathbb{M}^* \models \{\, \text{Sk}(\varphi) : \varphi \in \Gamma \,\}$. By the Löwenheim-Skolem theorem for first-order logic, $\{\, \text{Sk}(\varphi) : \varphi \in \Gamma \,\}$ has a model of cardinality κ, the reduct of which to the vocabulary of Γ is a structure $\mathbb{M}'$ of size κ such that $\mathbb{M}' \models^+ \Gamma$.   ⊣

---

[5] One can also use Gödel's first incompleteness theorem to prove Theorem 5.54. We are grateful to Antti Kuusisto for pointing out this simpler proof using Trakhtenbrot's theorem, which can be found in [17, pp. 171–172].

5.4.3 Separation

Craig's interpolation theorem [13] says that whenever φ is an FO_L sentence, and ψ is an $FO_{L'}$ sentence such that $\varphi \models \psi$, there is an $FO_{L \cap L'}$ sentence θ, called an *interpolant*, such that $\varphi \models \theta \models \psi$. We can use Craig's interpolation theorem to prove a separation theorem for IF logic.

Two IF formulas are *contrary* if there is no model in which both are true. If θ is a first-order formula, θ_{IF} denotes the IF formula for which θ is a shorthand.

Theorem 5.56 (Separation)   *Let φ and ψ be contrary IF_L sentences. Then there is an FO_L sentence θ such that*

$$\varphi \models^+ \theta_{IF} \quad and \quad \psi \models^+ \neg\theta_{IF}.$$

Proof   Since φ and ψ are contrary IF_L sentences, it follows that there is no model $\mathbb{M}$ such that $\mathbb{M} \models Sk(\varphi)$ and $\mathbb{M} \models Sk(\psi)$. Hence $Sk(\varphi) \models \neg Sk(\psi)$. We may assume the fresh function symbols introduced in $Sk(\varphi)$ and $Sk(\psi)$ are all distinct, so by Craig's interpolation theorem for first-order logic there is an FO_L sentence θ such that $Sk(\varphi) \models \theta \models \neg Sk(\psi)$. It follows that $Sk(\varphi) \models \theta$ and $Sk(\psi) \models \neg\theta$. Hence

$$\varphi \models^+ \theta_{IF} \quad and \quad \psi \models^+ \neg\theta_{IF}. \qquad \dashv$$

We can strengthen the separation theorem for IF logic by restricting attention to structures with at least two elements.

Theorem 5.57 (Burgess [8])   *Assume every structure has at least two elements, and let φ and ψ be contrary IF_L sentences. Then there is an IF_L sentence χ such that*

$$\varphi \equiv^+ \chi \quad and \quad \psi \equiv^+ \neg\chi.$$

Proof   Let φ_{MP} denote the Matching Pennies sentence

$$\forall x (\exists y/\{x\}) x = y,$$

which is neither true nor false in any structure. Let φ' be $\varphi \vee \varphi_{MP}$, and let ψ' be $\psi \vee \varphi_{MP}$. Observe that φ' and ψ' are contrary IF_L sentences such that $\varphi \equiv^+ \varphi'$ and $\psi \equiv^+ \psi'$. Also note that φ' and ψ' are never false because Eloise can always choose the disjunct φ_{MP}.

By the separation theorem for IF logic (Theorem 5.56), there is an FO_L sentence θ such that $\varphi' \models^+ \theta_{IF}$ and $\psi' \models^+ \neg\theta_{IF}$. The IF_L sentence χ we are looking for is $\varphi' \wedge (\neg\psi' \vee \theta_{IF})$. Observe that $\neg\chi$ is $\neg\varphi' \vee (\psi' \wedge \neg\theta_{IF})$

by definition. Furthermore,

$$\chi \equiv^+ \varphi' \wedge \theta_{\mathrm{IF}} \equiv^+ \varphi' \equiv^+ \varphi$$

because $\neg\psi'$ is never true, while

$$\neg\chi \equiv^+ \psi' \wedge \neg\theta_{\mathrm{IF}} \equiv^+ \psi' \equiv^+ \psi$$

because $\neg\varphi'$ is never true.   ⊣

Theorem 5.57 shows that knowing the class of structures in which an IF sentence is true reveals nothing about the class of structures in which it is false (beyond the fact that the two classes are disjoint).

5.4.4 Determinacy

Definition 5.58   An IF sentence φ is *determined in a structure* $\mathbb{M}$ if it is either true or false in $\mathbb{M}$. Otherwise φ is *undetermined in* $\mathbb{M}$. We say that φ is *determined* if it is determined in all suitable structures.   ⊣

Proposition 5.59 (Väänänen [66, Corollary 6.10])   *If φ is a determined IF sentence, then there is a first-order sentence θ such that φ is equivalent to θ_{IF}.*

Proof   The IF sentences φ and $\neg\varphi$ are contrary, so by the separation theorem for IF logic (Theorem 5.56) there is a first-order sentence θ such that $\varphi \models^+ \theta_{\mathrm{IF}}$ and $\varphi \models^- \theta_{\mathrm{IF}}$. If φ is determined, then $\theta_{\mathrm{IF}} \models^+ \varphi$ because

$$\mathbb{M} \models^+ \theta_{\mathrm{IF}} \quad \text{implies} \quad \mathbb{M} \not\models^- \theta_{\mathrm{IF}}$$
$$\text{implies} \quad \mathbb{M} \not\models^- \varphi$$
$$\text{implies} \quad \mathbb{M} \models^+ \varphi,$$

and a similar argument shows that $\theta_{\mathrm{IF}} \models^- \varphi$.   ⊣

Thus, an IF sentence that is not equivalent to a first-order sentence must be undetermined in some structure. It might also be interesting to know in which structure. It turns out that for certain IF sentences, this question can be given a more specific answer.

Theorem 5.60 (Väänänen [65])   *Let φ be an IF_L sentence, and let $\mathbb{M}$ be an infinite L-structure. Suppose that for every L-structure $\mathbb{M}'$ we have*

$$\mathbb{M}' \models^+ \varphi \quad \textit{iff} \quad \mathbb{M} \not\cong \mathbb{M}'.$$

Then φ is undetermined in exactly the structures isomorphic to $\mathbb{M}$.

Proof If M $\not\cong$ M', then M' $\models^+$ φ, and thus φ is determined in M'.

Conversely, if M $\cong$ M', then M' $\not\models^+$ φ. Suppose for the sake of a contradiction that M' $\models^-$ φ. Then M' $\models^+$ $\neg\varphi$. Given that M $\cong$ M' and M is infinite, M' must be infinite too. By the Löwenheim-Skolem theorem, there is a strictly larger structure M'' such that M'' $\models^+$ $\neg\varphi$. Whence M $\not\cong$ M'' and thus M'' $\models^+$ φ, which is impossible.   ⊣

To conclude the chapter, we consider an application of Theorem 5.60 to (Peano) arithmetic. The vocabulary $L = \{0, S\}$ of arithmetic consists of a constant symbol 0 and a unary function symbol S. In any model, the interpretation of the constant symbol 0 is called *zero*, and the interpretation of the function symbol S is called the *successor* function. The axioms of Peano arithmetic are meant to define what it means to be a *natural number*.

PA1. $\forall x\big(S(x) \neq 0\big)$.

PA2. $\forall x \forall y\big(S(x) = S(y) \rightarrow x = y\big)$.

PA3. $\forall X\Big[\big(X(0) \wedge \forall x[X(x) \rightarrow X\big(S(x)\big)]\big) \rightarrow \forall y X(y)\Big]$.

The first axiom asserts that zero is not the successor of any number. The second axiom asserts that the successor function is injective. Thus any model of the axioms must contain an infinite set of numbers

$$\{0, S(0), S\big(S(0)\big), \ldots\}.$$

The third axiom is called the induction axiom. It asserts that any set of natural numbers that includes zero and is closed under the successor function contains every natural number. Notice that the induction axiom is a second-order sentence because the variable X ranges over sets of natural numbers. We will discuss second-order logic in the next chapter.

Up to isomorphism, there is only one model of the Peano axioms, which is denoted $\mathbb{N}$.

Theorem 5.61 (Dedekind [15]) *Any two models of Peano arithmetic are isomorphic.*   ⊣

Example 5.62 Consider the IF sentence φ:

$$\forall x \exists y \forall u \big(\exists v/\{x, y\}\big)\big(\exists w/\{x, y, u, v\}\big)(\varphi_1 \wedge \varphi_2 \wedge \varphi_3 \wedge \varphi_4),$$

where

$$\varphi_1 \quad \text{is} \quad x = u \to y = v,$$
$$\varphi_2 \quad \text{is} \quad x = 0 \to y = 0,$$
$$\varphi_3 \quad \text{is} \quad (y = 0 \wedge u = S(x)) \to v = 0,$$
$$\varphi_4 \quad \text{is} \quad y = 0 \to x \neq w.$$

The Skolem form of φ is

$$\forall x \forall u \Big[(x = u \to f(x) = g(u)) \wedge (x = 0 \to f(x) = 0)$$
$$\wedge (f(x) = 0 \wedge u = S(x)) \to g(u) = 0$$
$$\wedge f(x) = 0 \to x \neq c \Big],$$

which is equivalent to

$$\forall x \Big[f(0) = 0$$
$$\wedge f(x) = 0 \to f(S(x)) = 0$$
$$\wedge f(x) = 0 \to x \neq c \Big].$$

Thus $M \models^+ \varphi$ if and only if there is an expansion of M with a unary function for which the preimage of 0^M includes 0^M and is closed under S^M, but is not the entire universe. In other words, φ is truth equivalent to the (contradictory) negation of PA3.

Let θ be the IF sentence

$$\exists x (S(x) = 0) \vee \exists x \exists y (x \neq y \wedge S(x) = S(y)) \vee \varphi.$$

Then for any $\{0, S\}$-structure M we have

$$M \models^+ \theta \quad \text{iff} \quad M \not\cong N.$$

By Theorem 5.60, the sentence θ is undetermined in exactly those $\{0, S\}$-structures that are isomorphic to N. ⊣

6

Expressive power of IF logic

In this chapter, we compare the expressivity of IF logic to that of other logical systems such as first-order logic and second-order logic. We will show that IF logic is a conservative extension of first-order logic in the sense that every property expressible by a first-order sentence can be expressed by an IF sentence. Furthermore, there is a class of structures defined by an IF sentence that is not first-order definable, which shows that IF logic is strictly more expressive than first-order logic. In fact, IF logic has exactly the same expressive power as the existential fragment of second-order logic.

At the end of the chapter we consider a fragment of IF logic. We will show that the fragment of IF logic consisting of sentences whose semantic games have perfect recall has the same expressive power as first-order logic.

6.1 Definability

One of the primary functions of a formal language is to enable us to express interesting properties. For example, the property of having two elements is captured by the first-order sentence

$$\exists x \exists y \big[x \neq y \wedge \forall z (x = z \vee y = z) \big]$$

because the sentence is true in every structure with two elements, and no others.

Definition 6.1 Let L be a vocabulary. A class **K** of L-structures is *defined* by an FO_L sentence φ if for every L-structure $\mathbb{M}$,

$$\mathbb{M} \in \mathbf{K} \quad \text{iff} \quad \mathbb{M} \models \varphi.$$

A class of L-structures is *first-order definable* (or *elementary*) if it is defined by an FO_L sentence. ⊣

Definition 6.2 A class **K** of L-structures is *defined* by an IF_L sentence φ if for every L-structure $\mathbb{M}$,

$$\mathbb{M} \in \mathbf{K} \quad \text{iff} \quad \mathbb{M} \models^+ \varphi.$$

A class of L-structures is IF *definable* if it is defined by an IF_L sentence. A pair $(\mathbf{K}^+, \mathbf{K}^-)$ of classes of L-structures is IF *definable* if there is an IF_L sentence φ such that for every L-structure $\mathbb{M}$ we have $\mathbb{M} \in \mathbf{K}^+$ if and only if $\mathbb{M} \models^+ \varphi$, and $\mathbb{M} \in \mathbf{K}^-$ if and only if $\mathbb{M} \models^- \varphi$. ⊣

Recall that every first-order formula φ can be viewed as a shorthand for an IF formula φ_{IF} in which every slash set is empty. The only difference between the semantic game for φ and the semantic game for φ_{IF} is the addition of the indistinguishability relations $\sim_\exists$ and $\sim_\forall$. Since no information is hidden from the players, however, the indistinguishability relations have no effect.

Theorem 6.3 *Let φ be an FO_L formula, and let φ_{IF} be the IF_L formula for which φ is a shorthand. Then for any suitable structure $\mathbb{M}$ and any assignment s whose domain contains* $\text{Free}(\varphi) = \text{Free}(\varphi_{IF})$,

$$\mathbb{M}, s \models \varphi \quad \text{iff} \quad \mathbb{M}, s \models^+ \varphi_{IF}.$$

If φ is a first-order sentence,

$$\mathbb{M} \models \varphi \quad \text{iff} \quad \mathbb{M} \models^+ \varphi_{IF}.$$

Proof By definition, any winning strategy for $G(\mathbb{M}, s, \varphi_{IF})$ is also a winning strategy for $G(\mathbb{M}, s, \varphi)$. Conversely, since every slash set in φ_{IF} is empty, two histories of $G(\mathbb{M}, s, \varphi_{IF})$ that lead to the same subformula are indistinguishable if and only if they induce identical assignments. By Proposition 3.15, if Eloise has a winning strategy for $G(\mathbb{M}, s, \varphi)$, then she has a memoryless winning strategy for $G(\mathbb{M}, s, \varphi)$, which is also a winning strategy for $G(\mathbb{M}, s, \varphi_{IF})$.

If φ is a sentence, then $\mathbb{M} \models \varphi$ if and only if $\mathbb{M}, \varnothing \models \varphi$ if and only if $\mathbb{M}, \{\varnothing\} \models^+ \varphi_{IF}$ if and only if $\mathbb{M} \models^+ \varphi_{IF}$. ⊣

Corollary 6.4 *Every elementary class of structures is IF definable.*

The converse of Corollary 6.4 does not hold. In Chapter 4, we exhibited an IF sentence φ_∞ that defines the class of structure with infinite universes (see Example 4.14). However, the class of infinite structures is not first-order definable, which shows that IF logic is strictly more expressive than first-order logic.

6.2 Second-order logic

One of the features of first-order logic that limits its expressive power is the fact that first-order quantifiers range over individuals, but not functions or relations. In order to quantify over functions and relations, we must ascend to second-order logic.

The syntax of second-order logic has two new sorts of variables: *relation variables* and *function variables* (collectively known as *second-order variables*). Like relation symbols and function symbols, each second-order variable comes with a specified arity. We assume that there are infinitely many relation and function variables of each arity. To distinguish ordinary first-order variables from relation variables and function variables, we will sometimes refer to them as *individual variables*.

The rules for generating second-order formulas and terms are the same as for first-order logic, plus additional rules for second-order variables.

Definition 6.5 Let L be a vocabulary. The set of *second-order L-terms* is generated by the finite application of the following rules:

- Every individual variable is a second-order L-term.
- Every constant symbol in L is a second-order L-term.
- If f is an n-ary function symbol in L and $t_1, \ldots, t_n$ are second-order L-terms, then $f(t_1, \ldots, t_n)$ is a second-order L-term.
- If F is an n-ary function variable and $t_1, \ldots, t_n$ are second-order L-terms, then $F(t_1, \ldots, t_n)$ is a second-order L-term. ⊣

Relation symbols can be combined with terms to form (atomic) formulas. Formulas can be combined with logical connectives and quantifiers to form (compound) formulas.

Definition 6.6 The second-order language generated by the vocabulary L, denoted SO_L, is generated by the finite application of the following rules:

- If t_1 and t_2 are second-order L-terms, then $(t_1 = t_2) \in \mathrm{SO}_L$.
- If R is an n-ary relation symbol in L and $t_1, \ldots, t_n$ are second-order L-terms, then $R(t_1, \ldots, t_n) \in \mathrm{SO}_L$.

- If X is an n-ary relation variable and $t_1, \ldots, t_n$ are second-order L-terms, then $X(t_1, \ldots, t_n) \in \mathrm{SO}_L$.
- If $\Phi \in \mathrm{SO}_L$, then $\neg\Phi \in \mathrm{SO}_L$.
- If $\Phi, \Phi' \in \mathrm{SO}_L$, then $(\Phi \vee \Phi') \in \mathrm{SO}_L$ and $(\Phi \wedge \Phi') \in \mathrm{SO}_L$.
- If $\Phi \in \mathrm{SO}_L$ and x is an individual variable, then $\exists x\Phi \in \mathrm{SO}_L$ and $\forall x\Phi \in \mathrm{SO}_L$.
- If $\Phi \in \mathrm{SO}_L$ and X is a relation variable, then $\exists X\Phi \in \mathrm{SO}_L$ and $\forall X\Phi \in \mathrm{SO}_L$.
- If $\Phi \in \mathrm{SO}_L$ and F is a function variable, then $\exists F\Phi \in \mathrm{SO}_L$ and $\forall F\Phi \in \mathrm{SO}_L$. ⊣

The elements of SO_L are called SO_L *formulas*. A *second-order formula* is an SO_L formula for some vocabulary L. As usual, when the vocabulary is irrelevant or clear from context we will not mention it explicitly.

An occurrence of a variable in a second-order formula is *free* if it is not bound by a quantifier of the appropriate type. If Φ is an atomic second-order formula, all its variables are free. For compound second-order formulas, we apply the clauses from Definition 3.5, plus:

$$\mathrm{Free}(QX\Phi) = \mathrm{Free}(\Phi) - \{X\},$$
$$\mathrm{Free}(QF\Phi) = \mathrm{Free}(\Phi) - \{F\},$$

where X is a relation variable, and F is a function variable.

The semantics of second-order logic is similar to that of first-order logic, except that a second-order assignment assigns values to three sorts of variables. The value of an individual variable must be an element of the universe, the value of an n-ary relation variable must be an n-ary relation on the universe, and the value of an n-ary function variable must be an n-ary function from the universe to itself.

Definition 6.7 Let Φ be a second-order formula, $\mathbb{M}$ a suitable structure, and s a second-order assignment. If X is an n-ary relation variable,

$$\mathbb{M}, s \models \exists X\Phi \quad \text{iff} \quad \mathbb{M}, s(X/R) \models \Phi, \text{ for some } R \subseteq M^n,$$
$$\mathbb{M}, s \models \forall X\Phi \quad \text{iff} \quad \mathbb{M}, s(X/R) \models \Phi, \text{ for every } R \subseteq M^n.$$

If f is an n-ary function variable,

$$\mathbb{M}, s \models \exists F\Phi \quad \text{iff} \quad \mathbb{M}, s(F/f) \models \Phi, \text{ for some } f \colon M^n \to M,$$
$$\mathbb{M}, s \models \forall F\Phi \quad \text{iff} \quad \mathbb{M}, s(F/f) \models \Phi, \text{ for every } f \colon M^n \to M. \quad ⊣$$

An n-ary function can be considered as an $(n+1)$-ary relation with the property that whenever the first n coordinates of any two $(n+1)$-tuples in the relation agree, their final coordinates also agree. Conversely, an n-ary relation R can be defined in terms of an n-ary function f by choosing a special element of the universe c, and defining

$$(x_1, \ldots, x_n) \in R \quad \text{iff} \quad f(x_1, \ldots, x_n) = c.$$

These considerations allow us to restrict our attention to a single kind of second-order variable. For our purposes, it will be useful to focus on function variables, and ignore relation variables.

We saw in Chapter 3 that if a first-order sentence φ is true in a model, then the Skolemization of φ is true in an expansion of the model. In Chapter 3, we extended the Skolemization procedure for first-order logic to IF logic. By using function variables instead of function symbols, we can define a second-order Skolemization procedure for IF logic that will allow us to identify second-order truth and falsity conditions for any IF sentence.

Definition 6.8   Let φ be an IF formula, and let U be a finite set of variables containing $\mathrm{Free}(\varphi)$. The *second-order Skolem form* (or *second-order Skolemization*) of $\psi \in \mathrm{Subf}(\varphi)$ with variables in U is defined exactly like the first-order Skolem form of ψ (Definition 4.9), except that each fresh function symbol $f_{(\exists x/W)\psi}$ is replaced by a corresponding function variable $F_{(\exists x/W)\psi}$. The *second-order Kreisel form* (or *second-order Kreiselization*) of ψ is defined analogously (see Definition 4.17).   ⊣

We will be a bit lax in our notation by allowing $\mathrm{Sk}_U(\psi)$ to denote either the first-order or the second-order Skolem form of ψ, depending on the context. Likewise, for $\mathrm{Kr}_U(\psi)$.

Definition 6.9   For any IF sentence φ, let φ^+ denote the second-order formula

$$\exists F_1 \ldots \exists F_m \, \mathrm{Sk}(\varphi),$$

where $F_1, \ldots, F_m$ are the free function variables in $\mathrm{Sk}(\varphi)$. Dually, let φ^- denote

$$\exists G_1 \ldots \exists G_n \, \mathrm{Kr}(\varphi),$$

where $G_1, \ldots, G_n$ are the free function variables in $\mathrm{Kr}(\varphi)$.   ⊣

For example, if φ_∞ is the IF sentence from Example 4.14, then φ_∞^+ is

$$\exists C \exists F \exists G \forall x \Big[G\big(F(x)\big) = x \land F(x) \neq C \Big],$$

which asserts that there exists an injection from the universe to itself whose range is not the entire universe, i.e., that the universe is (Dedekind) infinite. One can check that φ_∞^- is

$$\exists F \forall w \forall y \forall z \left[z = F(w) \wedge y \neq w \right],$$

which is tautologically false.

Theorem 6.10 *Let φ be an IF sentence and $\mathbb{M}$ a suitable structure. Then $\mathbb{M} \models^\pm \varphi$ if and only if $\mathbb{M} \models \varphi^\pm$.*

Proof $\mathbb{M} \models^+ \varphi$ if and only if there is an expansion $\mathbb{M}^*$ of $\mathbb{M}$ to the vocabulary $L^* = L \cup \{ f_\psi : \psi \in \text{Subf}_\exists(\varphi) \}$ such that $\mathbb{M}^* \models \text{Sk}(\varphi)$, and such an expansion exists if and only if $\mathbb{M} \models \varphi^+$.

Dually, $\mathbb{M} \models^- \varphi$ if and only if there is an expansion $\mathbb{M}^*$ of $\mathbb{M}$ to the vocabulary $L^* = L \cup \{ f_\psi : \psi \in \text{Subf}_\forall(\varphi) \}$ such that $\mathbb{M}^* \models \text{Kr}(\varphi)$, and such an expansion exists if and only if $\mathbb{M} \models \varphi^-$. $\dashv$

Thus we are justified in calling φ^+ the *second-order truth condition* of φ, and φ^- the *second-order falsity condition* of φ.

6.3 Existential second-order logic

By definition, φ^+ and φ^- consist of a block of existential second-order quantifiers followed by a second-order formula that does not contain any second-order quantifiers. Second-order formulas in which all second-order quantifiers are existential and appear at the front of the formula are called *existential second-order formulas*. The class of existential second-order formulas is denoted Σ_1^1 because of its position in the second-order quantifier-alternation hierarchy.

Definition 6.11 A second-order formula belongs to the class $\Sigma_0^1 = \Pi_0^1$ if it does not have any second-order quantifiers. It belongs to the class Σ_{n+1}^1 if it consists of a block of existential second-order quantifiers followed by a Π_n^1 formula, and to the class Π_{n+1}^1 if it consists of a block of universal second-order quantifiers followed by a Σ_n^1 formula. $\dashv$

Theorem 6.10 shows that every IF sentence has a Σ_1^1 truth condition and a Σ_1^1 falsity condition. The converse holds as well. Working in the context of branching quantifiers, Enderton [19] and Walkoe [67] independently proved that every Σ_1^1 sentence is equivalent to the second-order truth condition of an IF sentence. Furthermore, for every pair (Φ^+, Φ^-)

of contrary Σ_1^1 sentences there is an IF sentence φ such that φ^+ is equivalent to Φ^+, and φ^- is equivalent to Φ^- [8].

Enderton and Walkoe's result depends on the fact that every Σ_1^1 sentence is equivalent to a Σ_1^1 sentence with a specific form. First of all, we want the sentence to be in prenex normal form and free of first-order existential quantifiers.

Definition 6.12   A Σ_1^1 formula is in *Skolem normal form* if it has the form

$$\exists F_1 \ldots \exists F_n \forall x_1 \ldots \forall x_m \Psi$$

where $F_1, \ldots, F_n$ are function variables, $x_1, \ldots, x_m$ are individual variables, and Ψ is quantifier free. ⊣

Skolem [55] proved that every Σ_1^1 formula is equivalent to a formula in Skolem normal form.[1] The Σ_0^1 part of a Σ_1^1 sentence in Skolem normal form looks suspiciously like the second-order Skolem form of an IF sentence. However, the second-order Skolemization of an IF sentence cannot have second-order terms like $F(x, x, y)$ in which the same variable occurs as an argument multiple times. Nor can it have terms in which the same function variable occurs with different arguments, such as $F(x, y, z)$ and $F(z, y, x)$. If Ψ satisfies the further requirement that there are no second-order terms with nested function variables, e.g., $F\big(x, y, G(u, v)\big)$, then we can find an IF sentence φ such that $\mathrm{Sk}(\varphi)$ is equivalent to $\forall x_1 \ldots \forall x_m \Psi$.

Proposition 6.13   *Let Φ be a Σ_0^1 formula of the form $\forall x_1 \ldots \forall x_m \Psi$, where Ψ is a quantifier-free formula in negation normal form whose free individual variables are among $x_1, \ldots, x_m$. If for every second-order term $F(t_1, \ldots, t_k)$ in Ψ,*

(1) $\{t_1, \ldots, t_k\} \subseteq \{x_1, \ldots, x_m\}$,
(2) the terms $t_1, \ldots, t_k$ are distinct,
(3) for every second-order term $F'(t_1', \ldots, t_k')$ in Ψ, $F = F'$ implies that $t_i = t_i'$ for all $1 \le i \le k$,

then Φ is equivalent to the second-order Skolemization of an IF sentence, modulo renaming of function variables.

Proof   Suppose Ψ satisfies conditions (1)–(3). We will construct an IF sentence φ whose second-order Skolem form is equivalent to Φ modulo renaming function variables. Let φ_0 be Φ, and let ψ_0 be Ψ. Apply the following routine, starting from $i = 0$:

[1] English translations appear in [56, 57]. See [66, Theorem 6.12] for a nice proof.

- If $F(t_1, \ldots, t_k)$ is the left-most second-order term in φ_i:

$$\forall x_1 \ldots \forall x_m (\exists y_1 / W_1) \ldots (\exists y_i / W_i) \psi_i,$$

replace every instance of $F(t_1, \ldots, t_k)$ with a fresh individual variable y_{i+1} that is bound to an existential quantifier $(\exists y_{i+1} / W_{i+1})$, obtaining the formula φ_{i+1}:

$$\forall x_1 \ldots \forall x_m (\exists y_1 / W_1) \ldots (\exists y_i / W_i)(\exists y_{i+1} / W_{i+1}) \psi_{i+1},$$

where $W_{i+1} = \{x_1, \ldots, x_m, y_1, \ldots, y_i\} - \{t_1, \ldots, t_k\}$, and ψ_{i+1} is the quantifier-free formula such that ψ_i is

$$\text{Subst}(\psi_{i+1}, y_{i+1}, F(t_1, \ldots, t_k)).$$

By condition (3) every occurence of a given function variable F appears with the same arguments $t_1, \ldots, t_k$. Therefore ψ_{i+1} has no occurrences of F. Thus we create a sequence $\varphi_0, \ldots, \varphi_n$ of formulas, each with fewer function variables than the last. The formula φ_n contains no function variables; hence it is the desired IF formula φ. Since all the individual variables $x_1, \ldots, x_m$ were bound in the original formula, and each of the new variables $y_1, \ldots, y_n$ is bound by a new existential quantifier, we conclude that φ is an IF sentence.

We show by induction that for every $0 \leq i \leq n$, the second-order Skolemization of φ_i is equivalent to Φ modulo renaming function variables. The base case is easy because φ_0 is Φ by definition, and φ_0 is $\text{Sk}(\varphi_0)$ because φ_0 contains no existential quantifiers. For the inductive case, suppose $\text{Sk}(\varphi_i)$ is equivalent to Φ modulo renaming function variables. If we can show that ψ_i is equivalent to

$$\text{Sk}_{\{x_1, \ldots, x_m, y_1, \ldots, y_i\}}\big((\exists y_{i+1} / W_{i+1}) \psi_{i+1}\big)$$

modulo renaming function variables, then certainly $\text{Sk}(\varphi_i)$ is equivalent to $\text{Sk}(\varphi_{i+1})$, and we are done. Observe that

$$\{x_1, \ldots, x_m, y_1, \ldots, y_i\} - W_{i+1} = \{t_1, \ldots, t_k\}.$$

Hence $\text{Sk}_{\{x_1, \ldots, x_m, y_1, \ldots, y_i\}}\big((\exists y_{i+1} / W_{i+1}) \psi_{i+1}\big)$ is

$$\text{Subst}\big(\psi_{i+1}, y_{i+1}, F_{(\exists y_{i+1} / W_{i+1}) \psi_{i+1}}(t_1, \ldots, t_k)\big),$$

which is equivalent to ψ_i modulo renaming function variables. $\dashv$

Example 6.14 Consider the Σ_0^1 formula

$$\forall x \forall y R\big(x, F(x), y, G(y)\big).$$

Two iterations of the routine in the proof of Proposition 6.13 convert this formula into an IF sentence. First, we remove the second-order term $F(x)$ to obtain

$$\forall x \forall y \big(\exists v/\{y\}\big) R\big(x, v, y, G(y)\big).$$

Next we remove $G(y)$:

$$\forall x \forall y \big(\exists v/\{y\}\big)\big(\exists w/\{x, v\}\big) R(x, v, y, w).$$

We leave it as an exercise to check that Skolemizing the above formula yields the original formula.   ⊣

Example 6.15   Not every IF sentence has a second-order Skolem form that satisfies the conditions of Proposition 6.13. For instance, the formula

$$\forall x \forall y P\big(G(F(x))\big)$$

does not meet condition (1), even though it is the Skolemization of

$$\forall x \forall y \big(\exists z/\{y\}\big)\big(\exists u/\{x, y\}\big) P(u).$$ ⊣

The proof of the following theorem hinges on the fact that every Σ_0^1 formula can be massaged into an equivalent Σ_0^1 formula that does satisfy the conditions of Proposition 6.13.

Theorem 6.16 (Enderton [19] and Walkoe [67])   *For every Σ_1^1 sentence Φ there is an IF sentence φ in the same vocabulary such that for every structure* $\mathbb{M}$,

$$\mathbb{M} \models \Phi \quad \textit{iff} \quad \mathbb{M} \models^+ \varphi.$$

Proof   Let Φ be a Σ_1^1 sentence, which we may assume is of the form

$$\exists F_1 \ldots \exists F_n \forall x_1 \ldots \forall x_m \Psi$$

where Ψ is quantifier free. For each of the three conditions, we present an equivalent pair of Σ_0^1 formulas, so that by iteratively replacing the one by the other we obtain a formula that meets the condition at hand. The proof steps closely follow the proof of Theorem 6.15 from [66].

Condition (1). Let $R(t_1, \ldots, t_k)$ be an atomic subformula of Ψ such that $t_i = F(t_1', \ldots, t_\ell')$ and t_j' is a compound second-order term, for some $1 \leq i \leq k$ and $1 \leq j \leq \ell$. Let y be an individual variable that does not appear in Ψ. Then $R(t_1, \ldots, t_k)$ is equivalent to

$$\forall y \Big[y = t_j' \rightarrow$$
$$R(t_1, \ldots, t_{i-1}, F(t_1', \ldots, t_{j-1}', y, t_{j+1}', \ldots, t_\ell'), t_{i+1}, \ldots, t_k) \Big].$$

Thus we may assume that every second-order term in Ψ has the form $F(y_1, \ldots, y_\ell)$ where each y_j is an individual variable.

Condition (2). Let $R(t_1, \ldots, t_k)$ be an atomic subformula of Ψ such that $t_i = F(y_1, \ldots, y_\ell)$ is a second-order term in which the same variable occurs twice, say y_j and $y_{j'}$. Let z be an individual variable that does not appear in Ψ. Then $R(t_1, \ldots, t_k)$ is equivalent to

$$\forall z \Big[z = y_j \to$$
$$R\big(t_1, \ldots, t_{i-1}, F(y_1, \ldots, y_{j-1}, z, y_{j+1}, \ldots, y_\ell), t_{i+1}, \ldots, t_k\big) \Big].$$

Thus we may assume that the arguments of any second-order term $F(y_1, \ldots, y_\ell)$ that occurs in Ψ are all distinct variables.

Condition (3). Let $F(y_1, \ldots, y_k)$ and $F(z_1, \ldots, z_k)$ be second-order terms in Ψ that have the same function variable but different arguments, that is, for some $1 \leq i \leq k$, the individual variables y_i and z_i are not the same. Suppose that $\{y_1, \ldots, y_k\} \cap \{z_1, \ldots, z_k\} = \varnothing$. If this is not the case replace the variables in the intersection using the method for condition (2). Let F' be a k-ary function variable that does not appear in Φ. We replace F by F' in the second term, using the equivalence between $\forall x_1 \ldots \forall x_m \theta$ and

$$\exists F' \forall x_1 \ldots \forall x_m \Big[\big((y_1 = z_1 \wedge \ldots \wedge y_k = z_k) \to$$
$$F(y_1, \ldots, y_k) = F'(z_1, \ldots, z_k) \big) \wedge \theta' \Big],$$

where θ' is the result of replacing $F(z_1, \ldots, z_k)$ for $F'(z_1, \ldots, z_k)$ in θ wherever it appears. Thus we may assume that a given function variable F always appears with the same sequence of arguments.

Each modification preserves the earlier conditions. Thus Φ is equivalent to a Σ_1^1 sentence $\exists F_1 \ldots \exists F_{n'} \forall x_1 \ldots, x_{m'} \Psi'$, where Ψ' satisfies conditions (1)–(3). By Proposition 6.13, there is an IF sentence φ such that $\forall x_1 \ldots, x_{m'} \Psi'$ is equivalent to $\mathrm{Sk}(\varphi)$, hence Φ is equivalent to φ^+.   ⊣

Example 6.17   Consider the Σ_1^1 formula

$$\exists F \forall x \Big[x = F\big(F(x)\big) \wedge x \neq F(x) \Big],$$

which expresses evenness on finite structures (see Example 4.15). To meet condition (1), we replace the term $F(x)$ in $F\big(F(x)\big)$ by the new variable y:

$$\exists F \forall x \forall y \Big[y = F(x) \to \big(x = F(y) \wedge x \neq F(x) \big) \Big].$$

The resulting sentence violates condition (3) because the terms $F(x)$ and $F(y)$ have the same function variable but distinct arguments. Replacing the function variable in $F(y)$ with a new function variable G yields:

$$\exists F \exists G \forall x \forall y \Big[\big(x = y \rightarrow F(x) = G(y) \big)$$

$$\wedge \big(y = F(x) \rightarrow (x = G(y) \wedge x \neq F(x)) \big) \Big],$$

which is the second-order truth condition for

$$\forall x \forall y \big(\exists u / \{y\} \big) \big(\exists v / \{x, u\} \big) \Big[(x = y \rightarrow u = v)$$

$$\wedge \big(y = u \rightarrow (x = v \wedge x \neq u) \big) \Big]. \quad \dashv$$

Definition 6.18   A class of L-structures $\mathbf{K}$ is *defined* by a Σ_1^1 sentence Φ if for every L-structure $\mathbb{M}$,

$$\mathbb{M} \in \mathbf{K} \quad \text{iff} \quad \mathbb{M} \models \Phi.$$

A class of L-structures is Σ_1^1 *definable* (or *pseudo-elementary* if it is defined by a Σ_1^1 sentence. $\dashv$

Every elementary class is pseudo-elementary, but not vice versa. For example, the class of infinite structures is pseudo-elementary, but it is not elementary. It follows immediately from Theorem 6.16 that every pseudo-elementary class is IF definable.

By combining the compactness theorem for IF logic with Theorem 6.16, one can show that the class of finite structures is not pseudo-elementary. Thus the complement of a pseudo-elementary class is not necessarily pseudo-elementary, which implies that neither IF logic nor existential second-order logic is closed under contradictory negation.

Moreover, the orthogonality of the truth and falsity "coordinates" of an IF sentence implied by Theorem 5.57 means that the game-theoretic negation, which tells Eloise and Abelard to switch roles, does not correspond to any operation on classes of models.

Theorem 6.19 (Burgess [8])   *Excluding one-element structures, every pair of disjoint, pseudo-elementary classes is IF definable.*

Proof   Let $\mathbf{K}^+$ and $\mathbf{K}^-$ be pseudo-elementary classes. Then there are IF sentences φ and ψ such that $\mathbb{M} \in \mathbf{K}^+$ if and only if $\mathbb{M} \models^+ \varphi$, and $\mathbb{M} \in \mathbf{K}^-$ if and only if $\mathbb{M} \models^+ \psi$. Since $\mathbf{K}^+$ and $\mathbf{K}^-$ are disjoint, there is an IF sentence χ such that

$$\mathbb{M} \in \mathbf{K}^\pm \quad \text{iff} \quad \mathbb{M} \models^\pm \chi. \qquad \dashv$$

6.4 Perfect recall

Chapters 4 and 5 have given us a sense of the streams of information that flow through IF semantic games. We have seen how information can be hidden from the players, how the players can signal information to themselves, and how adjusting slash sets allows us to rearrange quantifiers.

Many of the information patterns specifiable by IF sentences cannot be obtained with first-order sentences. In the semantic game for an IF sentence, we have the ability to hide the value of any variable from any player at any time. In this section, we will consider IF sentences whose information flows satisfy a couple of natural conditions.

Informally, a player of an extensive game has *action recall* if he always remembers his own moves, and has *knowledge memory* if he never forgets information he once knew. A player with both action recall and knowledge memory is said to have *perfect recall* [35, 36].

In extensive games with perfect recall, ignorance of an opponent's move can be ascribed to "external" factors. For instance, we can imagine that the players write each move on a separate card, which is then placed face-up or face-down on a table. In the first case, the opponents learn which move was played and remember it for the rest of the game. In the second case, the opponents do not know which move was played unless and until the card is turned face up. In games with imperfect recall, a player's ignorance may be caused by "internal" factors such as forgetfulness or limited computational resources. Games that lack perfect recall have not received much attention in the game-theoretic literature, see [45, p. 204] and [47, p. 4].

We shall define perfect recall as a syntactic property of IF sentences. It is straightforward — but somewhat tedious — to prove that under the definition given below, an IF sentence has perfect recall for a player if and only if that player has perfect recall in all the sentence's semantic games [52].

For the purpose of illustration, consider the signaling sentence

$$\forall x \exists z (\exists y / \{x\}) x = y.$$

It was observed by van Benthem [4] that in this sentence Eloise does not have knowledge memory because she knows the value of x when she chooses the value of z, but not when she chooses the value of y. She does have action recall, because she knows the value she assigned to z when she picks y. As another example, Eloise has knowledge memory in the semantic game for the IF sentence

$$\forall x (\exists y/\{x\}) \exists z (\exists u/W) R(x,y,z,u), \qquad (6.1)$$

if and only if $W \subseteq \{z\}$ because she knows the values of x and y when she chooses the value of z. She does not have action recall if $y \in W$ or $z \in W$ because then she would have forgotten one of her own moves.

Does the observation that Eloise lacks perfect recall in certain IF sentences, such as the signaling sentence, render their games unplayable? Not necessarily. A well-known way to interpret games with imperfect recall is by regarding its players as coalitions of players whose interests are perfectly aligned. The popular card game bridge is played with four people divided into two teams. Each pair of partners must coordinate their moves, even though they are not allowed to see each other's cards. As the bidding moves around the table, each team "forgets" half of their cards. Thus the team as a whole does not have knowledge memory, even though each member does. The playability of IF games and their coalitional interpretation is discussed in [4, 28, 52], amongst others. See also the discussion of signaling games on pages 73–74.

To avoid complications that arise when a variable is assigned a value twice, we will restrict our attention to regular IF sentences. We introduce some auxiliary notions that help us talk about the set of variables that a player has seen so far in the game. For any regular IF sentence φ and any $\psi \in \mathrm{Subf}(\varphi)$, let $A_\varphi(\psi)$ denote the set of variables quantified superordinate to ψ. Note that if ψ is of the form $(Qx/W)\psi'$ then x is not contained in $A_\varphi(\psi)$. Now define

$$K_\varphi(\psi) = \begin{cases} A_\varphi(\psi) & \text{if } \psi \text{ is } \chi \circ \chi' \\ A_\varphi(\psi) - W & \text{if } \psi \text{ is } (Qx/W)\chi. \end{cases}$$

That is, $K_\varphi(\psi)$ is the set of variables the active player can see in the position corresponding to ψ.

Definition 6.20   For a regular IF sentence φ, Eloise has

- *action recall* if, whenever the variable x is existentially quantified superordinate to a subformula of the form $(\exists y/W)\psi$, we have

$$x \in K_\varphi((\exists y/W)\psi);$$

- *knowledge memory* if, whenever $\psi, \chi \in \mathrm{Subf}(\varphi)$ are subformulas that belong to Eloise and $\chi \in \mathrm{Subf}(\psi)$, we have $K_\varphi(\psi) \subseteq K_\varphi(\chi)$;
- *perfect recall* if she has both action recall and knowledge memory.

Abelard has action recall, knowledge memory, or perfect recall for φ if the dual clauses hold. ⊣

Example 6.21 Consider the IF sentence φ from (6.1). Let ψ be the subformula

$$(\exists u/W)R(x,y,z,u),$$

and observe that $K_\varphi(\psi) = \{x,y,z\} - W$. According to Definition 6.20, Eloise has action recall if both y and z belong to $K_\varphi(\psi)$, or (equivalently) if neither belong to W. Furthermore, Eloise has knowledge memory if

$$K_\varphi\big((\exists y/\{x\})\exists z\psi\big) \subseteq K_\varphi(\exists z\psi) \subseteq K_\varphi(\psi).$$

The first containment is automatic since the set on the left is empty. To obtain the second containment we need $\{x,y\} \subseteq K_\varphi(\psi)$, which is equivalent to $W \subseteq \{z\}$. ⊣

We will use the following equivalences.

Proposition 6.22 *Let φ and ψ be IF formulas, and let U be a finite set of variables that contains* Free(φ), Free(ψ), *and W. If x and y are distinct variables not in U,*

(a) $\forall x(\varphi \wedge \psi) \equiv_U \forall x\varphi \wedge \forall x\psi$,
    $\exists x(\varphi \vee \psi) \equiv_U \exists x\varphi \vee \exists x\psi$;

(b) $\forall x\forall y\varphi \equiv_U^+ \forall y\forall x\varphi$,
    $\exists x\exists y\varphi \equiv_U^- \exists y\exists x\varphi$;

(c) $\forall x(\exists y/W \cup \{x\})\varphi \equiv_U^+ (\exists y/W)\forall x\varphi$,
    $\exists x(\forall y/W \cup \{x\})\varphi \equiv_U^- (\forall y/W)\exists x\varphi$.

Proof Part (a) follows immediately from Proposition 5.23. To prove part (b), observe that by Proposition 5.29 and Lemma 5.42,

$$\forall x\forall y\varphi \equiv_U^+ \forall x(\forall y/\{x\})\varphi$$
$$\equiv_U \forall y(\forall x/\{y\})\varphi$$
$$\equiv_U^+ \forall y\forall x\varphi.$$

Part (c) is proved similarly. ⊣

It was shown by the third author [52, p. 44] that IF sentences have first-order truth conditions.

Theorem 6.23 *Every regular IF sentence for which Eloise (Abelard) has perfect recall is truth (falsity) equivalent to a first-order sentence.*

Proof Let φ be a regular IF sentence for which Eloise has perfect recall. Since we are only interested in truth equivalence, we may assume that the slash sets of universal quantifiers in φ are empty (see Lemma 5.42). If all the existential slash sets are empty, then by Theorem 6.3 we are done, so let us assume that $(\exists z/W)\chi$ is a subformula of φ such that the variable x belongs to W. Since Eloise has action recall, x must be universally quantified superordinate to $(\exists z/W)\chi$. Hence there must be a subformula of φ of the form $\forall x \psi$ such that $(\exists z/W)\chi$ is a subformula of ψ.

Since Eloise has knowledge memory, ψ cannot be of the form $\theta \vee \theta'$. If ψ is $\theta \wedge \theta'$ we can use Proposition 6.22(a) to distribute the universal quantifier over the conjunction to obtain $\forall x\theta \wedge \forall x\theta'$.

If $\forall x \psi$ is $\forall x \forall y \theta$, we can use Proposition 6.22(b) to obtain $\forall y \forall x \theta$. If it is $\forall x(\exists y/W')\theta$ we know that $x \in W'$ because Eloise has knowledge memory. Thus we can use Proposition 6.22(c) to pull the universal quantifier inside the existential quantifier to obtain

$$(\exists y/W' - \{x\})\forall x\theta.$$

Notice that each transformation preserves Eloise's perfect recall. By repeating these steps, we can eliminate all the variables in every slash set by pushing universal quantifiers as deep into the formula as necessary, but no further. Eventually, we will obtain an IF formula in which every slash set is empty that is truth equivalent to the original.

The proof for Abelard is similar.   ⊣

Corollary 6.24 *For any regular* IF *sentence whose truth (falsity) condition is not equivalent to a first-order sentence, Eloise (Abelard) does not have perfect recall.*

Theorem 6.23 does not imply that every regular IF sentence for which both Eloise and Abelard have perfect recall is equivalent to a first-order sentence, even though such sentences do have first-order truth and falsity conditions. For example, both players have perfect recall for the Matching Pennies sentence

$$\forall x(\exists y/\{x\})x = y.$$

When we push the universal quantifier inside the existential quantifier, we obtain the first-order truth condition $\exists y \forall x(x = y)$. To obtain the first-order falsity condition, we negate the formula and drop the slash set on the universal quantifier, which yields $\exists x \forall y(x \neq y)$.

Burgess showed that every pair of disjoint pseudo-elementary classes is IF definable (Theorem 6.19). Is it also the case that every pair of disjoint elementary classes can be defined by an IF sentence for which both players have perfect recall?

Theorem 6.25 *Excluding one-element structures, every pair of disjoint elementary classes is definable by a regular IF sentence for which both Eloise and Abelard have perfect recall.*

Proof Let $\mathbf{K}^+$ and $\mathbf{K}^-$ be disjoint elementary classes defined by φ and ψ, respectively. Let φ_{MP} denote the Matching Pennies sentence

$$\forall x (\exists y / \{x\}) x = y.$$

Let φ' be $\varphi_{IF} \vee \varphi_{MP}$ and ψ' be $\psi_{IF} \vee \varphi_{MP}$. Then φ' is an IF sentence that defines $\mathbf{K}^+$ but is never false, while ψ' defines $\mathbf{K}^-$ but is never false, either. By the separation theorem for IF logic (Theorem 5.56) there is a first-order sentence θ that is true in every $\mathbb{M} \in \mathbf{K}^+$ and false in every $\mathbb{M} \in \mathbf{K}^-$.

As in the proof of Theorem 5.57, the sentence we are looking for is

$$\varphi' \wedge (\neg \psi' \vee \theta_{IF}).$$

Observe that $\varphi' \wedge (\neg \psi' \vee \theta_{IF})$ is a regular IF sentence for which both Eloise and Abelard have perfect recall. $\dashv$

7

Probabilistic IF logic

Game-theoretic semantics attempts to characterize two logical notions, truth and falsity, in the setting of extensive game theory. We saw that in this framework, the truth (falsity) of an IF formula amounts to the existence of a winning strategy for Eloise (Abelard). Assuming that Abelard and Eloise act in their best interest, they will follow a winning strategy if such a strategy is available to either of them.

The question arises how Abelard and Eloise play if neither player has a winning strategy. Consider for instance the Matching Pennies sentence φ_{MP} that is defined by

$$\forall x (\exists y / \{x\}) x = y$$

and the Inverted Matching Pennies sentence φ_{IMP} that is defined by

$$\forall x (\exists y / \{x\}) x \neq y.$$

In the framework of game-theoretic semantics, the games of φ_{MP} and φ_{IMP} on any structure with at least two objects are undetermined: neither player has a winning strategy.

This does not mean that Eloise is indifferent as to which game she plays. Intuitively, if $\mathbb{M}$ has two objects, Eloise's odds of winning the game $G(\varphi_{\mathrm{MP}}, \mathbb{M})$ are equal to her odds of winning the game $G(\varphi_{\mathrm{IMP}}, \mathbb{M})$. But let's see what happens if $\mathbb{M}$ contains more than two objects. In the game of φ_{MP} on $\mathbb{M}$, given that Eloise is to choose the same object as Abelard, her chances of winning decrease. On the other side, in the game of φ_{IMP} on $\mathbb{M}$ her chances to win increase as the size of the $\mathbb{M}$ increases. So Eloise prefers playing $G(\mathbb{M}, \varphi_{\mathrm{IMP}})$ to playing $G(\mathbb{M}, \varphi_{\mathrm{MP}})$ if $\mathbb{M}$ contains more than two elements. The game-theoretic machinery laid out in Chapter 4 does not account for this observation as it does not explain how Abelard and Eloise play in undetermined games.

In this chapter we will introduce a new approach that allows Eloise and Abelard to randomize their strategies in such a way that the payoff resulting from their behavior is stable, i.e., is in *equilibrium*. We will see that applying the notion of equilibrium gives us a means to assign to IF games a value in the interval $[0,1]$. As a result, φ_{MP} has value $1/n$ on finite structures with n elements, whereas φ_{IMP} has value $(n-1)/n$.

There is an on-going debate in game theory about the interpretation of the notion of equilibrium and the implications it has on the behavior and resources of the players, etc. We are not going to enter this debate in this chapter. Nonetheless, the results we formulate are likely to provide interesting new ramifications for its articulation.

The bulk of the work presented in this chapter draws on recent results presented in [54]. The notion of applying equilibria to semantic games has been anticipated in [3, 7].

7.1 Equilibrium semantics

The idea to apply the minimax theorem to undetermined games (in the framework of Henkin quantifiers) goes back to Ajtai [7]. It was later taken up in [52] for a special class of quantifiers and generalized in [54] to IF logic. In this section we will interpret IF logic by strategic games and in this context we will use the minimax theorem to formulate a new semantic interpretation, called *equilibrium semantics*.

We first define the strategic counterpart $\Gamma(\mathbb{M}, s, \varphi)$ of an extensive game $G(\mathbb{M}, s, \varphi)$.

Definition 7.1 Let φ be an IF formula, $\mathbb{M}$ a suitable structure, and s an assignment in $\mathbb{M}$ whose domain contains $\mathrm{Free}(\varphi)$. Let G be the extensive game $G(\mathbb{M}, s, \varphi)$. The *strategic IF game* $\Gamma(\mathbb{M}, s, \varphi)$ is a strategic game with components defined as follows:

- N is $\{\exists, \forall\}$;
- S_p is the set of strategies of player p in G for each player $p \in N$;
- u_p is the utility function of player p such that

$$u_{\exists}(\sigma, \tau) = \begin{cases} 1 & \text{if playing } \sigma \text{ against } \tau \text{ is winning for Eloise in } G, \\ 0 & \text{if playing } \sigma \text{ against } \tau \text{ is losing for Eloise in } G, \end{cases}$$

$$u_{\forall}(\sigma, \tau) = 1 - u_{\exists}(\sigma, \tau).$$ ⊣

Whenever φ is a sentence and s is the empty assignment, we write $\Gamma(\mathbb{M}, \varphi)$ instead of $\Gamma(\mathbb{M}, s, \varphi)$. We shall often write S for $S_\exists$ and T for $S_\forall$, as well as u instead of $u_\exists$ and U instead of $U_\exists$.

Every strategic IF game is a one-sum game, since for every $\sigma \in S$ and $\tau \in T$, $u_\exists(\sigma, \tau) + u_\forall(\sigma, \tau) = 1$. We can transform every strategic IF game Γ into a zero-sum game Γ' whose utility function u_p' is defined from Γ's utility function u_p by $u_p'(\sigma, \tau) = 2u_p(\sigma, \tau) - 1$ for every $\sigma \in S$ and $\tau \in T$. By Proposition 2.17, every equilibrium in Γ' is an equilibrium in our strategic IF game Γ, and we can therefore apply the minimax theorem to strategic IF games.

Let $(\mu_1, \nu_1), \ldots, (\mu_i, \nu_i)$ enumerate the equilibria in a finite strategic IF game Γ. By the minimax theorem (Theorem 2.16), Γ has at least one equilibrium: $i \geq 1$. By Proposition 2.15, $U(\mu_1, \nu_1) = U(\mu_i, \nu_i)$ for all i. Thus, for any equilibrium (μ_i, ν_i), we can unambiguously refer to $U(\mu_i, \nu_i)$ as the *value* of Γ. We write $V(\Gamma)$ for the value of Γ. If $\Gamma = \Gamma(\mathbb{M}, s, \varphi)$, then we refer to $V(\Gamma)$ as the *value* of φ on $\mathbb{M}$ relative to s. It is obvious that strategic IF games take values in the interval $[0, 1]$.

Any strategic IF game $\Gamma(\mathbb{M}, s, \varphi)$ is finite if and only if the structure $\mathbb{M}$ is finite. Therefore, it is not guaranteed by the minimax theorem that $\Gamma(\mathbb{M}, s, \varphi)$ has an equilibrium if $\mathbb{M}$ is infinite. In this chapter we shall only deal with finite strategic IF games.

Example 7.2 Let $\mathbb{M}$ be a finite structure with universe $\{1, \ldots, n\}$. Recall our earlier example φ_{MP}, the Matching Pennies sentence

$$\forall x (\exists y / \{x\}) x = y.$$

We have seen that the Skolem form of φ_{MP} is $\forall x (x = c)$ and that the Kreisel form of φ_{MP} is $\forall y (d \neq y)$. This means that in $\Gamma(\mathbb{M}, \varphi_{\mathrm{MP}})$ every pure strategy σ_i instructs Eloise to pick the object $i \in M$, and similarly for strategies τ_j of Abelard. Let us write $S = \{\sigma_1, \ldots, \sigma_n\}$ for the pure strategies of Eloise in $\Gamma(\mathbb{M}, \varphi_{\mathrm{MP}})$, and $T = \{\tau_1, \ldots, \tau_n\}$ for the pure strategies of Abelard. The utility functions are given by

$$u_\exists(\sigma_i, \tau_j) = \begin{cases} 1 & \text{if } i = j, \\ 0 & \text{if } i \neq j, \end{cases}$$

$$u_\forall(\sigma_i, \tau_j) = 1 - u_\exists(\sigma_i, \tau_j).$$

Eloise's utility function is displayed as a payoff matrix in Figure 7.1(a). Letting μ^* denote the uniform strategy over S and ν^* the uniform strategy over T, we claim that (μ^*, ν^*) is an equilibrium in $\Gamma(\mathbb{M}, \varphi_{\mathrm{MP}})$. For any pure strategy $\sigma_i \in S$, $U_\exists(\sigma_i, \nu^*) = \sum_{\tau_j} \nu^*(\tau_j) u_\exists(\sigma_i, \tau_j)$. Eloise's

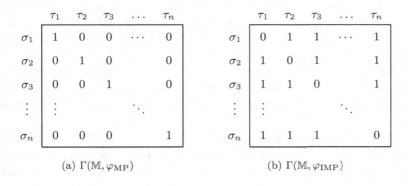

	τ_1	τ_2	τ_3	$\cdots$	τ_n
σ_1	1	0	0	$\cdots$	0
σ_2	0	1	0		0
σ_3	0	0	1		0
$\vdots$	$\vdots$			$\ddots$	
σ_n	0	0	0		1

(a) $\Gamma(\mathbb{M}, \varphi_{\mathrm{MP}})$

	τ_1	τ_2	τ_3	$\cdots$	τ_n
σ_1	0	1	1	$\cdots$	1
σ_2	1	0	1		1
σ_3	1	1	0		1
$\vdots$	$\vdots$			$\ddots$	
σ_n	1	1	1		0

(b) $\Gamma(\mathbb{M}, \varphi_{\mathrm{IMP}})$

Figure 7.1 The payoff matrices of Eloise in Matching Pennies and Inverted Matching Pennies

utility function $u_\exists$ returns 1 if $i = j$; otherwise it returns 0. Hence, $U_\exists(\sigma_i, \nu^*) = \nu^*(\tau_i) = 1/n$. Similar reasoning shows that for each $\tau_j \in T$, $U_\forall(\mu^*, \tau_j) = (n-1)/n$. By Proposition 2.18, (μ^*, ν^*) is an equilibrium. We conclude that the value of $\Gamma(\mathbb{M}, \varphi_{\mathrm{MP}})$ is $1/n$. ⊣

Example 7.3 Let $\mathbb{M}$ be the structure from the previous example and φ_{IMP} the Inverted Matching Pennies sentence $\forall x (\exists y/\{x\}) x \neq y$. In $\Gamma(\mathbb{M}, \varphi_{\mathrm{IMP}})$, the set of strategies of Eloise and Abelard are the same as in the game $\Gamma(\mathbb{M}, \varphi_{\mathrm{MP}})$. The utility function of Eloise in $\Gamma(\mathbb{M}, \varphi_{\mathrm{IMP}})$ is the inverse of the utility function of $\Gamma(\mathbb{M}, \varphi_{\mathrm{MP}})$, as shown in Figure 7.1(b).

The two uniform strategies μ^* and ν^* from the previous example are also an equilibrium in this case. However, in this game they yield an expected utility for Eloise of $(n-1)/n$, that is, the value of $\Gamma(\mathbb{M}, \varphi_{\mathrm{IMP}})$ is $(n-1)/n$. ⊣

The following result compares the values of strategic IF games to the truth values of extensive IF games.

Proposition 7.4 *Let φ be an IF formula, $\mathbb{M}$ a suitable structure, and s an assignment in $\mathbb{M}$ whose domain contains $\mathrm{Free}(\varphi)$. Let $G = G(\mathbb{M}, s, \varphi)$ and $\Gamma = \Gamma(\mathbb{M}, s, \varphi)$. Then*

- *Eloise has a winning strategy in G if and only if the value of Γ is 1;*
- *Abelard has a winning strategy in G if and only if the value of Γ is 0.*

Proof We prove the first claim. Let σ be a winning strategy in G, that is, $u(\sigma, \tau) = 1$ for every strategy $\tau \in T$. Consequently, for each mixed strategy $\nu \in \Delta(T)$, $U(\sigma, \nu) = 1$. Let $\mu \in \Delta(S)$ be the degenerate mixed strategy that assigns probability 1 to σ. We have $U(\mu, \nu) = 1$. Hence, condition (1) of Proposition 2.18 is met. Condition (2) follows from the fact that $u(\sigma, \tau) = 1$ for every strategy $\tau \in T$. Conditions (3) and (4) are immediately satisfied since $U(\mu, \nu) = 1$ is the maximal value that can be secured in Γ.

For the converse direction, suppose (μ, ν) is an equilibrium in Γ with value 1. Let $\sigma \in S$ be a strategy in the support of μ. By condition (1) of Proposition 2.18, $U(\sigma, \nu) = U(\mu, \nu) = 1$, that is, σ is winning against every strategy τ in the support in ν. For the pure strategies τ that are not in the support of ν, we derive from condition (4) of Proposition 2.18 that $U(\mu, \tau) \geq 1$. Since the maximal value in Γ is 1, this reduces to $U(\mu, \tau) = 1$. We conclude that σ is a winning strategy in G since $u(\sigma, \tau) = 1$, for every $\tau \in T$. ⊣

Observe that by the direction from right-to-left of the proof of Proposition 7.4, every strategy σ in the support of μ is winning. Consequently, (σ, ν) is an equilibrium in Γ for every σ in the support of μ. This shows that randomizing over winning strategies does not improve the expected utility.

Proposition 7.4 shows how the satisfaction of an IF formula under game-theoretic semantics is connected to the value of strategic IF games. We will now introduce a new satisfaction relation that is based on the values of strategic IF games.

Definition 7.5 Let φ be an IF formula, $\mathbb{M}$ a suitable structure, and s an assignment in $\mathbb{M}$ whose domain contains $\text{Free}(\varphi)$. Let $0 \leq \varepsilon \leq 1$ and $\Gamma = \Gamma(\mathbb{M}, s, \varphi)$. We define the relation "the value of φ in $\mathbb{M}$ relative to s is ε under *equilibrium semantics*," written $\models_{\text{Eq}}^{\varepsilon}$, by the following clause:

$$\mathbb{M}, s \models_{\text{Eq}}^{\varepsilon} \varphi \quad \text{iff} \quad V\big(\Gamma(\mathbb{M}, s, \varphi)\big) = \varepsilon.$$

In the same vein we introduce the following two relations:

$$\mathbb{M}, s \models_{\text{Eq}}^{\geq \varepsilon} \varphi \quad \text{iff} \quad V\big(\Gamma(\mathbb{M}, s, \varphi)\big) > \varepsilon$$

$$\mathbb{M}, s \models_{\text{Eq}}^{\leq \varepsilon} \varphi \quad \text{iff} \quad V\big(\Gamma(\mathbb{M}, s, \varphi)\big) < \varepsilon. \qquad \dashv$$

It follows from the analyses in Examples 7.2 and 7.3 that on finite structures $\mathbb{M}$ with n objects,

$$\mathbb{M} \models_{\text{Eq}}^{1/n} \varphi_{\text{MP}} \quad \text{and} \quad \mathbb{M} \models_{\text{Eq}}^{(n-1)/n} \varphi_{\text{IMP}}.$$

Equilibrium semantics is a conservative extension of game-theoretic semantics.

Corollary 7.6 *Let φ be an* IF *formula,* $\mathbb{M}$ *a suitable structure, and s an assignment in $\mathbb{M}$ whose domain contains* Free(φ). *Then*

$$\mathbb{M}, s \models^+ \varphi \quad \textit{iff} \quad \mathbb{M}, s \models^1_{Eq} \varphi,$$
$$\mathbb{M}, s \models^- \varphi \quad \textit{iff} \quad \mathbb{M}, s \models^0_{Eq} \varphi.$$

Proof Immediate from Proposition 7.4 and Definition 7.5. $\dashv$

We can study other model-theoretic notions of IF logic under equilibrium semantics. Analogous to Definitions 6.1 and 6.2 we say that a class **K** of L-structures is *ε-defined under equilibrium semantics* by an IF$_L$ sentence φ if for every L-structure $\mathbb{M}$,

$$\mathbb{M} \in \mathbf{K} \quad \text{iff} \quad \mathbb{M} \models^\varepsilon_{Eq} \varphi.$$

A class **K** of L-structures is *ε-definable under equilibrium semantics* if it is ε-defined under equilibrium semantics by some IF$_L$ sentence.

In the same vein we can define that **K** is *<ε-definable under equilibrium semantics* and *>ε-definable under equilibrium semantics* with respect to the relations $\models^{<\varepsilon}_{Eq}$ and $\models^{>\varepsilon}_{Eq}$, respectively.

Corollary 7.6 shows that every IF definable class is 1-definable under equilibrium semantics. The expressive power of IF logic under equilibrium semantics was studied in [54].

Example 7.7 Under equilibrium semantics IF formulas are assigned values in the interval $[0, 1]$. We can use the new semantics to describe the proportion of objects in the universe that satisfy a certain property φ. As an example we will show how the *Rescher quantifier* R can be defined in IF logic under the new interpretation. The Rescher quantifier R$x\varphi$ says that more than half of the objects satisfy φ:

$$\mathbb{M}, s \models \mathrm{R}x\varphi \quad \text{iff} \quad 2\Big|\big\{a \in M : \mathbb{M}, s(x/a) \models \varphi\big\}\Big| > |M|.$$

The Rescher quantifier R$xP(x)$ can be defined in IF logic under equilibrium semantics. Let $\mathbb{M}$ be a finite structure with universe $\{0, \ldots, n-1\}$ that interprets the binary function $(\ldots + \ldots) \bmod n$ as usual on the integers. Note that $(i + j) \bmod n$ is an element of M, for each $i, j \in M$.

Let φ be the IF sentence

$$\exists x \big(\forall y/\{x\}\big) P\big((x + y) \bmod n\big).$$

Suppose $\mathbb{M}$ is a structure with universe $\{0, 1, 2, 3, 4\}$ and $P^\mathbb{M} = \{0, 1, 2\}$.

	τ_0	τ_1	τ_2	τ_3	τ_4
σ_0	1	1	1	0	0
σ_1	0	1	1	1	0
σ_2	0	0	1	1	1
σ_3	1	0	0	1	1
σ_4	1	1	0	0	1

Figure 7.2 The payoff matrix of Eloise in the game $\Gamma(\mathbb{M}, \varphi)$

The payoff matrix of Eloise in $\Gamma = \Gamma(\mathbb{M}, \varphi)$ is given in Figure 7.2. Since more than half of the objects in M are P objects we have $\mathbb{M} \models RxP(x)$.

Let $S = \{\sigma_0, \ldots, \sigma_{n-1}\}$ and $T = \{\tau_0, \ldots, \tau_{n-1}\}$ and let μ and ν be the uniform mixed strategies with support S and T, respectively. Then

$$U(\mu, \nu) = \sum_{\sigma_i \in S} \sum_{\tau_j \in T} \mu(\sigma_i)\nu(\tau_j)u(\sigma_i, \tau_j) = 5 \left[\sum_{\tau_j \in M} \frac{1}{5} \cdot \frac{1}{5} u(\sigma_k, \tau_j) \right] = \frac{3}{5},$$

for an arbitrary $\sigma_k \in S$. With the help of Proposition 2.18 we can easily show that $U(\mu', \nu) = U(\mu, \nu') = 3/5$ for all $\mu' \in \Delta(S)$ and $\nu' \in \Delta(T)$. It follows that (μ, ν) is an equilibrium in Γ.

To generalize this argument, let n and $P^{\mathbb{M}}$ be arbitrary. Then the value of $\Gamma(\mathbb{M}, \varphi)$ equals m/n, where m is the cardinality of $P^{\mathbb{M}}$. Thus we arrive at the following characterization of the Rescher quantifier:

$$\mathbb{M} \models RxP(x) \quad \text{iff} \quad \mathbb{M} \models_{\text{Eq}}^{>1/2} \varphi. \qquad \dashv$$

7.2 Monotonicity rules

Strategic IF games can be large objects. In this and the following section we develop a toolkit to reduce strategic IF games to smaller games that have the same value.

In this section we will give a set of rules that tell us how to compute the value of certain IF formulas in terms of the values of its parts. We will show that the value of an IF formula $\varphi \vee \psi$ is the maximum of the values of φ and ψ. The plan of the proof is as follows. We first express the pure strategies in a strategic game $\Gamma = \Gamma(\mathbb{M}, s, \varphi \vee \psi)$ in a way that reflects their mode of composition from the pure strategies in $\Gamma_\varphi = \Gamma(\mathbb{M}, s, \varphi)$ and $\Gamma_\psi = \Gamma(\mathbb{M}, s, \psi)$. Then we show that every mixed strategy μ in Γ defines two mixed strategies $\mu_{|\varphi}$ and $\mu_{|\psi}$ in Γ_φ and Γ_ψ, respectively, such that the expected utility of μ is a weighted sum of the expected utilities of $\mu_{|\varphi}$ and $\mu_{|\psi}$. Finally, we will show that the equilibria in Γ correspond to the equilibria in Γ_φ and Γ_ψ, which will prove our claim.

Next we will show how the value of some existentially quantified formulas $(\exists x/W)\varphi$ is the maximum of the values of φ relative to all possible assignments to x. The structure of the proof is similar to the previous one. Finally we show that the values of IF formulas φ and $\neg\varphi$ add up to 1.

7.2.1 Constructing strategies for connective games

Consider the IF formula $\varphi \vee \psi$. Let $\mathbb{M}$ be a suitable structure and s an assignment in $\mathbb{M}$ whose domain contains $\text{Free}(\varphi \vee \psi)$.

The game $\Gamma = \Gamma(\mathbb{M}, s, \varphi \vee \psi)$ on the one hand and the games Γ_φ and Γ_ψ on the other hand are closely related. Every history $(s, \varphi, a_1, \ldots, a_m)$ for $G_\varphi = G(\mathbb{M}, s, \varphi)$ corresponds to a history $(s, \varphi \vee \psi, \varphi, a_1, \ldots, a_m)$ for $G = G(\mathbb{M}, s, \varphi \vee \psi)$. The converse holds as well: every history for G corresponds to a history for either G_φ or G_ψ in the same sense as above. We will exploit this insight by showing that there is a natural way to identify the set of strategies of player p in Γ with the product of his or her strategies in Γ_φ and Γ_ψ.

Let σ_φ and σ_ψ be pure strategies of Eloise in Γ_φ and Γ_ψ, respectively. Let $\chi \in \{\varphi, \psi\}$ be a choice for the disjunction. We write $\oplus(\chi, \sigma_\varphi, \sigma_\psi)$ for the function whose arguments are histories of Eloise in G such that

$$\oplus(\chi, \sigma_\varphi, \sigma_\psi)(s, \varphi \vee \psi) = \chi,$$

$$\oplus(\chi, \sigma_\varphi, \sigma_\psi)(s, \varphi \vee \psi, \chi, \ldots) = \begin{cases} \sigma_\varphi(s, \varphi, \ldots) & \text{if } \chi \text{ is } \varphi, \\ \sigma_\psi(s, \psi, \ldots) & \text{if } \chi \text{ is } \psi. \end{cases}$$

To check that $\oplus(\chi, \sigma_\varphi, \sigma_\psi)$ is a pure strategy of Eloise in Γ, we need to verify that it meets the uniformity requirements imposed by the subformulas θ in $\varphi \vee \psi$ of the form $(\exists x/W)\theta'$. To this end, consider two histories $h, h' \in H_\theta$ from G such that $h \approx_W h'$. Let $\chi \in \{\varphi, \psi\}$ be the disjunction that contains θ as a subformula. Let g and g' be the histories from $G(\mathbb{M}, s, \chi)$ that correspond to h and h'. Since the assignments induced by h and h' coincide with those induced by g and g', respectively, it follows that $g \approx_W g'$. Since σ_χ is a strategy in Γ_χ, we have $\sigma_\chi(g) = \sigma_\chi(g')$.

By the construction of $\oplus(\chi, \sigma_\varphi, \sigma_\psi)$, $\oplus(\chi, \sigma_\varphi, \sigma_\psi)(h) = \oplus(\chi, \sigma_\varphi, \sigma_\psi)(h')$. Hence for every pure strategy σ_φ in Γ_φ and σ_ψ in Γ_ψ, their composition $\oplus(\chi, \sigma_\varphi, \sigma_\psi)$ is a pure strategy in Γ.

We introduce similar notions for Abelard's pure strategies in Γ, Γ_φ and Γ_ψ. If τ_φ and τ_ψ are pure strategies of Abelard in Γ_φ and Γ_ψ, respectively, then $\oplus(\tau_\varphi, \tau_\psi)$ denotes the function whose arguments are histories of Abelard in G such that

$$\oplus(\tau_\varphi, \tau_\psi)(s, \varphi \vee \psi, \chi, \ldots) = \begin{cases} \tau_\varphi(s, \varphi, \ldots) & \text{if } \chi \text{ is } \varphi, \\ \tau_\psi(s, \psi, \ldots) & \text{if } \chi \text{ is } \psi. \end{cases}$$

It can be checked by a similar argument to the one above that $\oplus(\tau_\varphi, \tau_\psi)$ is a strategy of Abelard in Γ.

Let

$$\bigoplus(\{\varphi, \psi\}, \Gamma_\varphi, \Gamma_\psi) = (\{\exists, \forall\}, \{S_\oplus, T_\oplus\}, u_\oplus)$$

be the strategic game such that

$$S_\oplus = \{\oplus(\chi, \sigma_\varphi, \sigma_\psi) \colon \chi \in \{\varphi, \psi\}, \sigma_\varphi \in S_\varphi, \sigma_\psi \in S_\psi\},$$
$$T_\oplus = \{\oplus(\tau_\varphi, \tau_\psi) \colon \tau_\varphi \in T_\varphi, \tau_\psi \in T_\psi\},$$

where S_φ is the set of Eloise's pure strategies in Γ_φ etc., and

$$u_\oplus\big(\oplus(\chi, \sigma_\varphi, \sigma_\psi), \oplus(\tau_\varphi, \tau_\psi)\big) = \begin{cases} u_\varphi(\sigma_\varphi, \tau_\varphi) & \text{if } \chi \text{ is } \varphi, \\ u_\psi(\sigma_\psi, \tau_\psi) & \text{if } \chi \text{ is } \psi. \end{cases}$$

Proposition 7.8 Γ *and* $\bigoplus(\{\varphi, \psi\}, \Gamma_\varphi, \Gamma_\psi)$ *are identical.*

Proof We have already seen that every strategy in $S_\oplus$ and $T_\oplus$ is a strategy of Eloise and Abelard in Γ, respectively. For the converse direction, let σ be a strategy of Eloise in Γ. For any $\theta \in \{\varphi, \psi\}$, let σ_θ be the function whose arguments are the histories $g = (s, \theta, \ldots)$ of Eloise in G_θ such that $\sigma_\theta(g) = \sigma(s, \varphi \vee \psi, \theta, \ldots)$. In order to prove that σ_θ is a strategy in Γ_θ, let $g \approx_W g'$ be two histories in G_θ. Let h and h' be the histories in G that correspond to g and g', respectively. Since h and g correspond, they induce the same assignment. The same is true for h' and g'. Hence $h \approx_W h'$ and therefore $\sigma(h) = \sigma(h')$. By the construction of σ_θ, $\sigma_\theta(g) = \sigma_\theta(g')$. It follows that σ and $\oplus(\chi, \sigma_\varphi, \sigma_\psi)$ agree on all histories in G, where $\chi \in \{\varphi, \psi\}$ is the disjunct chosen by σ.

The case for strategies τ of Abelard is similar.

It is straightforward to see that for every $\sigma = \oplus(\chi, \sigma_\varphi, \sigma_\psi)$ and $\tau = \oplus(\tau_\varphi, \tau_\psi)$, $u(\sigma, \tau) = u_\oplus\big(\oplus(\chi, \sigma_\varphi, \sigma_\psi), \oplus(\tau_\varphi, \tau_\psi)\big)$.   ⊣

The same analysis can be applied to IF formulas of the form $\varphi \wedge \psi$.

7.2.2 Probability theory for mixed strategies

Let P be a probability distribution. A *random variable* A is a variable that ranges over a set of values $\{a_1, \ldots, a_n\}$ that it can assume. The *prior probability* that variable A assumes value a is denoted by $P(A = a)$. If the random variable A is clear from the context, we shall simply write $P(a)$ for $P(A = a)$. Let $A_1, \ldots, A_n$ be random variables and let $a_1, \ldots, a_n$ be values that they can assume, respectively. Then

$$P(A_1 = a_1, \ldots, A_n = a_n)$$

denotes the probability that $A_1 = a_1, \ldots, A_{n-1} = a_{n-1}$, and $A_n = a_n$.

According to this notation the probability that μ_p assigns to σ_p is written as $\mu_p(X_p = \sigma_p)$, where X_p is the random variable that ranges over the pure strategies of player p. The shorthand notation allows us to use the familiar notation $\mu_p(\sigma_p)$ for the same probability.

Let $\Gamma_\oplus = \bigoplus(\{\varphi, \psi\}, \Gamma_\varphi, \Gamma_\psi)$. Let $X_\vee$ be the random variable that ranges over $\{\varphi, \psi\}$ and X_χ the random variable that ranges over Eloise's strategies in Γ_χ for each $\chi \in \{\varphi, \psi\}$. If μ is a mixed strategy of Eloise in $\Gamma_\oplus$, then $\mu(X_\vee = \chi, X_\varphi = \sigma_\varphi, X_\psi = \sigma_\psi) = \mu(\chi, \sigma_\varphi, \sigma_\psi)$ denotes the probability $\mu(\oplus(\chi, \sigma_\varphi, \sigma_\psi))$. Symmetrically, when Y_χ ranges over Abelard's strategies in Γ_χ and ν is a mixed strategy in Abelard in $\Gamma_\oplus$, then $\nu(Y_\varphi = \tau_\varphi, Y_\psi = \tau_\psi) = \nu(\tau_\varphi, \tau_\psi)$ denotes the probability $\nu(\oplus(\tau_\varphi, \tau_\psi))$.

The expression

$$P(A_1 = a_1, \ldots, A_i = a_i \mid A_{i+1} = a_{i+1}, \ldots, A_n = a_n)$$

denotes the *posterior probability* that $A_1 = a_1, \ldots, A_i = a_i$ given that $A_{i+1} = a_{i+1}, \ldots, A_n = a_n$. Again if the random variables are clear from the context, we shall simply write $P(a_1, \ldots, a_n)$ and

$$P(a_1, \ldots, a_i \mid a_{i+1}, \ldots, a_n).$$

The posterior probability $P(a_1, \ldots, a_i \mid a_{i+1}, \ldots, a_n)$ can be defined in terms of prior probabilities:

$$P(a_1, \ldots, a_i \mid a_{i+1}, \ldots, a_n) = \frac{P(a_1, \ldots, a_n)}{P(a_{i+1}, \ldots, a_n)}. \tag{7.1}$$

If the events $A_{i+1} = a_{i+1}, \ldots, A_n = a_n$ are all independent, then

$$P(a_{i+1}, \ldots, a_n) = P(a_{i+1}) \cdots P(a_n),$$

in which case we can multiply both sides of (7.1) by $P(a_{i+1})$ to obtain the following equation, known as the *product rule*:

$$P(a_{i+1})P(a_1, \ldots, a_i \mid a_{i+1}, \ldots, a_n) = P(a_1, \ldots, a_{i+1} \mid a_{i+2}, \ldots, a_n).$$

Summing over the values a_i that the random variable A_i can take eliminates the random variable A_i from $P(a_1, \ldots, a_i \mid a_{i+1}, \ldots a_n)$:

$$\sum_{a_i} P(a_1, \ldots, a_i \mid a_{i+1}, \ldots a_n) = P(a_1, \ldots, a_{i-1} \mid a_{i+1}, \ldots a_n).$$

Applying this rule we get

$$\mu(\chi) = \sum_{\sigma_\varphi} \sum_{\sigma_\psi} \mu(\chi, \sigma_\varphi, \sigma_\psi),$$

$$\mu(\sigma_\varphi \mid \chi) = \sum_{\sigma_\psi} \mu(\sigma_\varphi, \sigma_\psi \mid \chi),$$

and by the product rule,

$$\mu(\chi, \sigma_\varphi, \sigma_\psi) = \mu(\chi)\mu(\sigma_\varphi, \sigma_\psi \mid \chi).$$

Observe that our terminology allows us to use the symbol μ both as a unary function (e.g. $\mu(\varphi)$) and as a binary function (e.g. $\mu(\sigma_\varphi \mid \varphi)$). All these functions are derived from the ternary probability distribution μ on $\{\varphi, \psi\} \times S_\oplus \times T_\oplus$. The same holds for Abelard's probability distribution ν that can be used as a unary function $\nu(\tau_\varphi)$.

If B is a random variable that can take the value b, then $P_{\mid b}$ denotes the probability distribution such that

$$P_{\mid b}(a_1, \ldots, a_i \mid a_{i+1}, \ldots a_n) = P(a_1, \ldots, a_i \mid a_{i+1}, \ldots a_n, b).$$

For instance, $\mu_{\mid \varphi}(\sigma_\varphi, \sigma_\psi) = \mu(\sigma_\varphi, \sigma_\psi \mid \varphi)$.

Example 7.9 Let $X_\vee$ range over $\{\varphi, \psi\}$, X_φ over $A = \{a_1, a_2\}$ and X_ψ over $B = \{b_1, b_2\}$. Let μ be the probability distribution over $\{\varphi, \psi\} \times A \times B$ as defined in Figure 7.3. Note that μ corresponds to a mixed strategy of Eloise in $\bigoplus(\{\varphi, \psi\}, \Gamma_\varphi, \Gamma_\psi)$, where $\Gamma_\varphi = \Gamma(\mathbb{M}, \exists x P(x))$ and $\Gamma_\psi = \Gamma(\mathbb{M}, \exists x Q(x))$ for some suitable two-element structure $\mathbb{M}$.

By elimination of the random variable X_ψ,

$$\mu(\varphi, a_1) = \sum_{b_j \in B} \mu(\varphi, a_1, b_j) = \frac{1}{8} + \frac{1}{16} = \frac{3}{16}.$$

Similarly,

$$\mu(\varphi) = \sum_{a_i \in A} \sum_{b_j \in B} \mu(\varphi, a_i, b_j) = \frac{1}{8} + \frac{1}{16} + \frac{1}{4} + \frac{5}{16} = \frac{3}{4}.$$

$X_\vee$	X_φ	X_ψ	μ
φ	a_1	b_1	$1/8$
φ	a_1	b_2	$1/16$
φ	a_2	b_1	$1/4$
φ	a_2	b_2	$5/16$
ψ	a_1	b_1	$1/8$
ψ	a_1	b_2	$1/16$
ψ	a_2	b_1	$1/32$
ψ	a_2	b_2	$1/32$

Figure 7.3 The probability distribution μ

and $\mu(\psi) = 1/4$. In addition, $\mu(a_1) = 3/8$. Observe that $\mu(\varphi, a_1)$ is distinct from $\mu(\varphi)\mu(a_1)$.

Posterior probabilities can be computed with the product rule:

$$\mu(a_1 \mid \varphi) = \sum_{b_j \in B} \mu(a_1, b_j \mid \varphi) = \sum_{b_j \in B} \frac{\mu(\varphi, a_1, b_j)}{\mu(\varphi)} = \frac{1}{4}$$

and $\mu(a_1 \mid \psi) = 7/8$. Observe that $\mu(\varphi, a_1)$ is distinct from $\mu(a_1 \mid \varphi)$.

The probability distribution $\mu_{|\varphi}$ is the mixed strategy in Γ_φ such that $\mu_{|\varphi}(a_1) = 1/4$ and $\mu_{|\varphi}(a_2) = 3/4$. ⊣

From a mixed strategy in the game $\Gamma_\oplus$ one can extract mixed strategies for both players in Γ_φ and Γ_ψ.

Proposition 7.10 *Let* $\Gamma_\oplus = \bigoplus(\{\varphi, \psi\}, \Gamma_\varphi, \Gamma_\psi)$. *Let* μ *be a mixed strategy of Eloise in* $\Gamma_\oplus$ *and* ν *a mixed strategy of Abelard in* $\Gamma_\oplus$. *Then*

- $\mu_{|\chi}$ *is a mixed strategy of Eloise in* Γ_χ *for each* $\chi \in \{\varphi, \psi\}$;
- ν *is a mixed strategy of Abelard in both* Γ_φ *and* Γ_ψ.

Proof

- Let χ be φ. We have

$$\sum_{\sigma_\varphi} \mu_{|\varphi}(\sigma_\varphi) = \sum_{\sigma_\varphi} \sum_{\sigma_\psi} \mu(\sigma_\varphi, \sigma_\psi \mid \varphi) = \sum_{\sigma_\varphi} \sum_{\sigma_\psi} \frac{\mu(\varphi, \sigma_\varphi, \sigma_\psi)}{\mu(\varphi)},$$

which is equal to

$$\frac{1}{\mu(\varphi)} \sum_{\sigma_\varphi} \sum_{\sigma_\psi} \mu(\varphi, \sigma_\varphi, \sigma_\psi) = 1.$$

The case for ψ is similar.

- We have

$$\sum_{\tau_\varphi} \nu(\tau_\varphi) = \sum_{\tau_\varphi} \sum_{\tau_\psi} \nu(\tau_\varphi, \tau_\psi) = 1.$$

The case for ψ is similar. ⊣

7.2.3 Monotonicity rule for connectives

A mixed strategy μ of Eloise in Γ encompasses the mixed strategy $\mu_{|\varphi}$ in Γ_φ and the mixed strategy $\mu_{|\psi}$ in Γ_ψ. We show that the expected utility of playing μ in Γ is the weighted sum of the expected utilities of playing $\mu_{|\varphi}$ in Γ_φ and playing $\mu_{|\psi}$ in Γ_φ.

Lemma 7.11 *Let $\varphi \vee \psi$ be an IF formula, $\mathbb{M}$ a suitable structure and s an assignment in $\mathbb{M}$ whose domain contains $\mathrm{Free}(\varphi \vee \psi)$. Let $U_\oplus$ be the expected utility function of $\bigoplus(\{\varphi, \psi\}, \Gamma_\varphi, \Gamma_\psi)$ and U_χ the expected utility function of Γ_χ for each $\chi \in \{\varphi, \psi\}$. Then*

$$U_\oplus(\mu, \nu) = \mu(\varphi)U_\varphi(\mu_{|\varphi}, \nu) + \mu(\psi)U_\psi(\mu_{|\psi}, \nu).$$

Proof First we observe that $U_\oplus(\mu, \nu) = k_\varphi + k_\psi$, where

$$k_\chi = \sum_{\sigma_\varphi} \sum_{\sigma_\psi} \sum_{\tau_\varphi} \sum_{\tau_\psi} \mu(\chi, \sigma_\varphi, \sigma_\psi)\nu(\tau_\varphi, \tau_\psi)u_\oplus\big(\oplus(\chi, \sigma_\varphi, \sigma_\psi), \oplus(\tau_\varphi, \tau_\psi)\big).$$

We can reduce $u_\oplus$ in k_χ to the utility function u_χ of the game Γ_χ:

$$u_\oplus\big(\oplus(\chi, \sigma_\varphi, \sigma_\psi), \oplus(\tau_\varphi, \tau_\psi)\big) = u_\chi(\sigma_\chi, \tau_\chi).$$

By the product rule, $\mu(\chi, \sigma_\varphi, \sigma_\psi) = \mu(\chi)\mu(\sigma_\varphi, \sigma_\psi \mid \chi)$. Suppose χ is φ. Since we sum over all σ_ψ in k_φ, we can eliminate the random variable X_ψ from $\mu(\sigma_\varphi, \sigma_\psi \mid \varphi)$:

$$k_\varphi = \sum_{\sigma_\varphi} \sum_{\tau_\varphi} \sum_{\tau_\psi} \mu(\varphi)\mu(\sigma_\varphi \mid \varphi)\nu(\tau_\varphi, \tau_\psi)u_\varphi(\sigma_\varphi, \tau_\varphi).$$

Similarly we can eliminate the random variable Y_ψ from $\nu(\tau_\varphi, \tau_\psi)$:

$$k_\varphi = \sum_{\sigma_\varphi} \sum_{\tau_\varphi} \mu(\varphi)\mu(\sigma_\varphi \mid \varphi)\nu(\tau_\varphi)u_\varphi(\sigma_\varphi, \tau_\varphi) = \mu(\varphi)U_\varphi(\mu_{|\varphi}, \nu).$$

In the same way it follows that $k_\psi = \mu(\psi)U_\psi(\mu_{|\psi}, \nu)$. ⊣

It follows from the previous result that there is no mixed strategy μ whose expected utility in Γ is greater than that of $\mu_{|\varphi}$ in Γ_φ and $\mu_{|\psi}$ in Γ_ψ, that is, expected utility of the mixed strategy μ is *monotone* with respect to expected utilities of the mixed strategies $\mu_{|\varphi}$ and $\mu_{|\psi}$. We will use this insight when we prove the monotonicity rule for the connectives.

Theorem 7.12 *Let φ and ψ be IF formulas, $\mathbb{M}$ a suitable structure and s an assignment in $\mathbb{M}$ whose domain contains* Free$(\varphi) \cup$ Free(ψ). *Then*

$$V\big(\Gamma(\mathbb{M}, s, \varphi \vee \psi)\big) = \max\big(V(\Gamma_\varphi), V(\Gamma_\psi)\big),$$
$$V\big(\Gamma(\mathbb{M}, s, \varphi \wedge \psi)\big) = \min\big(V(\Gamma_\varphi), V(\Gamma_\psi)\big).$$

Proof Fix $\Gamma_\oplus = \bigoplus(\{\varphi, \psi\}, \Gamma_\varphi, \Gamma_\psi)$. Write v_φ for $V(\Gamma_\varphi)$ and v_ψ for $V(\Gamma_\psi)$. We may assume without loss of generality that $v_\varphi = \max(v_\varphi, v_\psi)$. Let $(\mu_\varphi^*, \nu_\varphi^*)$ be an equilibrium in Γ_φ and (μ_ψ^*, ν_ψ^*) an equilibrium in Γ_ψ.

Fix any of Eloise's pure strategies σ_ψ' in Γ_ψ. Let $\hat{\mu} \in \Delta(S_\oplus)$ be the mixed strategy in $\Gamma_\oplus$ such that for every pure strategy $\sigma = \oplus(\chi, \sigma_\varphi, \sigma_\psi)$,

$$\hat{\mu}(\sigma) = \begin{cases} \mu_\varphi^*(\sigma_\varphi) & \text{if } \chi \text{ is } \varphi \text{ and } \sigma_\psi = \sigma_\psi', \\ 0 & \text{otherwise.} \end{cases}$$

Let $\hat{\nu}$ be the mixed strategy in $\Gamma_\oplus$ such that for every pure strategy $\tau = \oplus(\tau_\varphi, \tau_\psi)$,

$$\hat{\nu}(\tau) = \nu_\varphi^*(\tau_\varphi)\nu_\psi^*(\tau_\psi).$$

Observe that $\hat{\nu}(\tau_\chi) = \nu_\chi^*(\tau_\chi)$ for every pure strategy τ_χ of Abelard in Γ_χ.

It suffices to prove that (i) $U(\hat{\mu}, \hat{\nu}) = v_\varphi$ and (ii) $(\hat{\mu}, \hat{\nu})$ is an equilibrium in $\Gamma_\oplus$.

(i) Let $S_\oplus^\varphi$ be the set of strategies in $S_\oplus$ that choose disjunct φ. Then

$$U_\oplus(\hat{\mu}, \hat{\nu}) = \sum_{\sigma \in S_\oplus^\varphi} \sum_{\tau \in T_\oplus} \hat{\mu}(\sigma)\hat{\nu}(\tau)u_\oplus(\sigma, \tau)$$

$$= \sum_{\sigma_\varphi} \sum_{\tau_\varphi} \sum_{\tau_\psi} \hat{\mu}(\varphi, \sigma_\varphi, \sigma_\psi')\hat{\nu}(\tau_\varphi, \tau_\psi)u_\varphi(\sigma_\varphi, \tau_\varphi),$$

where σ_φ ranges over the pure strategies of Eloise in Γ_φ, etc. Eliminating the random variable Y_ψ yields:

$$\sum_{\sigma_\varphi} \sum_{\tau_\varphi} \hat{\mu}(\varphi, \sigma_\varphi, \sigma_\psi')\hat{\nu}(\tau_\varphi)u_\varphi(\sigma_\varphi, \tau_\varphi),$$

By construction, $\hat{\mu}(\varphi, \sigma_\varphi, \sigma'_\psi) = \mu^*_\varphi(\sigma_\varphi)$ and $\hat{\nu}(\tau_\varphi) = \nu^*_\varphi(\tau_\varphi)$. Hence, $U_\oplus(\hat{\mu}, \hat{\nu}) = U_\varphi(\mu^*_\varphi, \nu^*_\varphi) = v_\varphi$.

(ii) Suppose $\mu \in \Delta(S_\oplus)$ is a mixed strategy such that $U_\oplus(\mu, \hat{\nu}) > v_\varphi$. By Lemma 7.11, $U_\oplus(\mu, \hat{\nu}) = \mu(\varphi)U_\varphi(\mu_{|\varphi}, \hat{\nu}) + \mu(\psi)U_\psi(\mu_{|\psi}, \hat{\nu})$. Since $\mu(\varphi)$ and $\mu(\psi)$ are values between 0 and 1 that sum up to 1, there is a $\chi \in \{\varphi, \psi\}$ such that $U_\chi(\mu_{|\chi}, \hat{\nu}) > v_\varphi$. Suppose χ is φ. By a similar reasoning as above, $U_\varphi(\mu_{|\varphi}, \hat{\nu}) = U_\varphi(\mu_{|\varphi}, \nu^*_\varphi)$. Since $(\mu^*_\varphi, \nu^*_\varphi)$ is an equilibrium in Γ_φ, we must have that $U_\varphi(\mu_{|\varphi}, \nu^*_\varphi) \leq v_\varphi$. Contradiction.

Suppose χ is ψ. We have $U_\psi(\mu_{|\psi}, \hat{\nu}) = U_\psi(\mu_{|\psi}, \nu^*_\psi)$. Since (μ^*_ψ, ν^*_ψ) is an equilibrium in Γ_ψ, $U_\psi(\mu_{|\psi}, \nu^*_\psi) \leq v_\psi$. Recall that $v_\psi \leq v_\varphi$ and a contradiction follows.

In a similar way it can be shown that a contradiction follows from the assertion that Abelard has a strategy ν such that $U_\oplus(\hat{\mu}, \nu) < v_\varphi$.

The case for $\varphi \wedge \psi$ is left as an exercise to the reader. ⊣

This result allows us to consider a connective $\varphi \circ \psi$ as a choice point between playing Γ_φ and Γ_ψ. The player associated with $\circ$ compares his or her expected utilities in Γ_φ and Γ_ψ, and chooses the game with highest expected utility.

7.2.4 Constructing strategies for quantifier games

We explored the correspondence between the strategies in $\Gamma(\mathbb{M}, s, \varphi \vee \psi)$ on the one hand and the strategies in $\Gamma(\mathbb{M}, s, \varphi)$ and $\Gamma(\mathbb{M}, s, \psi)$ on the other hand. In a similar way there exists a correspondence between the strategies in $\Gamma(\mathbb{M}, s, (\exists x/W)\varphi)$ and the strategies in the games $\Gamma(\mathbb{M}, s(x/a_1), \varphi), \ldots, \Gamma(\mathbb{M}, s(x/a_n), \varphi)$, where $a_1, \ldots, a_n$ enumerate the objects in $\mathbb{M}$. We will first introduce the composition operator $\oplus$ for strategic IF games associated with existentially quantified formulas.

In the context of a singleton assignment team, the existential quantifier $(\exists x/W)\varphi$ can be conceived of as a big disjunction $\bigvee_{a \in M} \varphi$ over the objects a in the structure at hand. We can exploit this analogy between existential quantifiers and disjunctions to prove the quantifier monotonicity rules. The structure of the proof is similar to that of the proof of the monotonicity rules for the connectives. We need to account for one essential difference between connectives and quantifiers, though. Once Eloise has chosen a disjunct in $\varphi \vee \psi$, her choice is known to both players throughout the game. However, it may happen that one of the players does not know the move made for the outer quantifier in the IF formula $(\exists x/W)\varphi$.

To restore the analogy we will only consider IF formulas $(\exists x/W)\varphi$ in which both players can see the value assigned to x throughout the game. Note that it is not required that the value assigned to x by $(\exists x/W)$ is not overwritten in the remainder of the game. We say the variable x is *public* in the IF formula $(Qx/W)\varphi$ if $x \in K_\varphi(\psi)$ for every subformula ψ in φ. For instance, x is public in $\exists x\exists x\forall y R(x,y)$ but not in

$$\exists x(\exists y/\{x\})R(x,y) \quad \text{and} \quad \exists x(\forall y/\{x\})R(x,y).$$

Fix a suitable structure $\mathbb{M}$ with universe $\{a_1,\ldots,a_n\}$ and an assignment s in $\mathbb{M}$ whose domain contains $\text{Free}((\exists x/W)\varphi)$. As usual we let $G = G(\mathbb{M}, s, (\exists x/W)\varphi)$ and for every $a \in M$, we denote by G_a the extensive game $G(\mathbb{M}, s(x/a), \varphi)$. Similarly we write Γ for $\Gamma(\mathbb{M}, s, (\exists x/W)\varphi)$ and Γ_a for $\Gamma(\mathbb{M}, s(x/a), \varphi)$.

For each $a_i \in M$, we let σ_{a_i} range over the pure strategies for Eloise in Γ_{a_i}. For each $b \in M$, we let $\oplus(b, \sigma_{a_1}, \ldots, \sigma_{a_n})$ denote the function whose arguments are histories of Eloise in G such that

$$\oplus(b, \sigma_{a_1}, \ldots, \sigma_{a_n})(s, (\exists x/W)\varphi) = (x,b)$$
$$\oplus(b, \sigma_{a_1}, \ldots, \sigma_{a_n})(s, (\exists x/W)\varphi, (x,b), \ldots) = \sigma_b(s(x/b), \varphi, \ldots).$$

To see that $\oplus(b, \sigma_{a_1}, \ldots, \sigma_{a_n})$ is a strategy of Eloise in G, let $h \approx_U h'$ be two histories from G belonging to Eloise. The histories h and h' are both of the form $(s, (\exists x/W)\varphi, (x,b), \ldots)$. Since x is public, $x \notin U$. It follows that the assignments induced by h and h' agree on x. Let g and g' be the histories of the form $(s(x/b), \varphi, \ldots)$ from G_b that correspond to h and h', respectively.

Suppose h is of the form

$$(s, (\exists x/W)\varphi, (x,b), \ldots, (x,c), \ldots), \tag{7.2}$$

that is, the value assigned to x by $(\exists x/W)$ is overwritten at least once. Since g corresponds to h, g is of the form

$$(s, \varphi, (x,b), \ldots, (x,c), \ldots).$$

It follows that the assignments induced by g and h coincide. The same is true for the assignments induced by g' and h'.

Suppose h is not of the form (7.2), that is, the value of x is never overwritten. Then the assignments induced by g and h assign b to x, and they also agree on every other variable in their domain. Again, we have that the same is true for the assignments induced by g' and h'.

We conclude that $g \approx_U g'$. Since σ_b is a strategy in G_b, $\sigma_b(g) = \sigma(g')$. Then by construction, $\oplus(b, \sigma_{a_1}, \ldots, \sigma_{a_n})(h) = \oplus(b, \sigma_{a_1}, \ldots, \sigma_{a_n})(h')$ and the claim follows.

For each $a_i \in M$, we let τ_{a_i} range over the pure strategies of Abelard in Γ_{a_i}. By $\oplus(\tau_{a_1}, \ldots, \tau_{a_n})$ we denote the function whose arguments are histories of Abelard in G such that

$$\oplus(\tau_{a_1}, \ldots, \tau_{a_n})\big(s, (\exists x/W)\varphi, (x, b), \ldots\big) = \tau_b\big(s(x/b), \varphi, \ldots\big).$$

The function $\oplus(\tau_{a_1}, \ldots, \tau_{a_n})$ is in fact a strategy of Abelard in G. The proof of this claim is analogous to the earlier claim that $\oplus(b, \sigma_{a_1}, \ldots, \sigma_{a_n})$ is a strategy of Eloise in G.

Example 7.13 If x fails to be public in $(\exists x/W)\varphi$, there are composite functions $\oplus(\ldots)$ that are not strategies. For instance, let φ be the IF sentence $\exists x(\forall y/\{x\})R(x, y)$. Fix a structure $\mathbb{M}$ with at least two elements a, b. For $c \in \{a, b\}$, fix

$$\Gamma_c = \Gamma\Big(\mathbb{M}, \{(x, c)\}, (\forall y/\{x\})R(x, y)\Big).$$

Let τ_a be Abelard's pure strategy from Γ_a that picks a and τ_b his pure strategy from Γ_b that picks b. Consider the function $\oplus(\tau_a, \tau_b)$, which assigns (y, a) to $h = (\varnothing, \varphi, (x, a))$ and (y, b) to $h' = (\varnothing, \varphi, (x, b))$. Since $h \approx_{\{x\}} h'$, $\oplus(\tau_a, \tau_b)$ is not a strategy of Abelard in the game of φ.

The same argument applies to the IF sentence φ' in which Eloise moves twice: $\exists x(\exists y/\{x\})R(x, y)$. Let σ_a be Eloise's pure strategy from Γ_a that picks a and σ_b her pure strategy from Γ_b that picks b. Consider the function $\oplus(a, \sigma_a, \sigma_b)$. Since $h \approx_{\{x\}} h'$, $\oplus(a, \sigma_a, \sigma_b)$ is not a strategy of Eloise in the game of φ. The fact that h' does not materialize if Eloise assigns a to x is irrelevant in this context.   ⊣

Let

$$\bigoplus(M, \Gamma_{a_1}, \ldots, \Gamma_{a_n}) = (\{\exists, \forall\}, \{S_\oplus, T_\oplus\}, u_\oplus)$$

be the strategic game such that

$$S_\oplus = \big\{\oplus(b, \sigma_{a_1}, \ldots, \sigma_{a_n}) : b \in M, \sigma_{a_1} \in S_{a_1}, \ldots, \sigma_{a_n} \in S_{a_n}\big\},$$
$$T_\oplus = \big\{\oplus(\tau_{a_1}, \ldots, \tau_{a_n}) : \tau_{a_1} \in T_{a_1}, \ldots, \tau_{a_n} \in T_{a_n}\big\},$$

and

$$u_\oplus\big(\oplus(b, \sigma_{a_1}, \ldots, \sigma_{a_n}), \oplus(\tau_{a_1}, \ldots, \tau_{a_n})\big) = u_b(\sigma_b, \tau_b),$$

where u_b is the utility function of the strategic IF game Γ_b.

Proposition 7.14 Γ *and* $\bigoplus(M, \Gamma_{a_1}, \ldots, \Gamma_{a_n})$ *are identical.*

Proof Analogous to the proof of Proposition 7.8. $\dashv$

Let $X_{\exists}$ be the random variable that ranges over the objects in M, X_{a_i} the random variable that ranges over the strategies in S_{a_i}, and Y_{a_j} the random variable that ranges over the strategies in T_{a_j}. Fix a mixed strategy $\mu \in \Delta(S_{\oplus})$ from $\Gamma_{\oplus} = \bigoplus(M, \Gamma_{a_1}, \ldots, \Gamma_{a_n})$. Then for any pure strategy $\oplus(b, \sigma_{a_1}, \ldots, \sigma_{a_n})$ in $S_{\oplus}$,

$$\mu(X_{\exists} = b, X_{a_1} = \sigma_{a_1}, \ldots, X_{a_n} = \sigma_{a_n}) = \mu(b, \sigma_{a_1}, \ldots, \sigma_{a_n})$$

denotes the probability $\mu(\oplus(b, \sigma_{a_1}, \ldots, \sigma_{a_n}))$. For any of Abelard's pure strategies $\oplus(\tau_{a_1}, \ldots, \tau_{a_n})$ in $T_{\oplus}$,

$$\nu(Y_{a_1} = \tau_{a_1}, \ldots, Y_{a_n} = \tau_{a_n}) = \nu(\tau_{a_1}, \ldots, \tau_{a_n})$$

denotes the probability $\nu(\oplus(\tau_{a_1}, \ldots, \tau_{a_n}))$. We will use the notation for probability distributions introduced earlier by which, for instance,

$$\mu(b) = \sum_{\sigma_{a_1}} \cdots \sum_{\sigma_{a_n}} \mu(b, \sigma_{a_1}, \ldots, \sigma_{a_n}),$$

$$\mu_{|b}(\sigma_b) = \mu(\sigma_b \mid b).$$

Proposition 7.15 *Let* $\Gamma_{\oplus} = \bigoplus(M, \Gamma_{a_1}, \ldots, \Gamma_{a_n})$. *Let* μ *be a mixed strategy of Eloise in* $\Gamma_{\oplus}$ *and* ν *a mixed strategy of Abelard in* $\Gamma_{\oplus}$. *Then*

- $\mu_{|b}$ *is a mixed strategy of Eloise in* Γ_b *for each* $b \in M$;
- ν *is a mixed strategy of Abelard in all* $\Gamma_{a_1}, \ldots, \Gamma_{a_n}$.

Proof Analogous to the proof of Proposition 7.10. $\dashv$

7.2.5 Monotonicity rule for quantifiers

We proceed to show the quantifier analogues of Lemma 7.11 and Theorem 7.12.

Lemma 7.16 *Let* $(\exists x/W)\varphi$ *be an* IF *formula in which* x *is public. Let* $\mathbb{M}$ *be a suitable structure and* s *an assignment in* $\mathbb{M}$ *whose domain contains* $\mathrm{Free}((\exists x/W)\varphi)$. *Let* U_{a_i} *be the expected utility function of* Γ_{a_i} *for any* $a_i \in M$, *and* $U_{\oplus}$ *the expected utility function of* $\bigoplus(M, \Gamma_{a_1}, \ldots, \Gamma_{a_n})$. *Then*

$$U_{\oplus}(\mu, \nu) = \sum_{b \in M} \mu(b) U_b(\mu_{|b}, \nu).$$

Proof   First we observe that $U_\oplus(\mu, \nu) = \sum_{b \in M} k_b$, where

$$k_b = \sum_{\sigma_{a_1}} \cdots \sum_{\sigma_{a_n}} \sum_{\tau_{a_1}} \cdots \sum_{\tau_{a_n}} \mu(b, \sigma_{a_1}, \ldots, \sigma_{a_n}) \nu(\tau_{a_1}, \ldots, \tau_{a_n}) u_\oplus(\sigma_\oplus, \tau_\oplus),$$

$\sigma_\oplus = \oplus(b, \sigma_{a_1}, \ldots, \sigma_{a_n})$ and $\tau_\oplus = \oplus(\tau_{a_1}, \ldots, \tau_{a_n})$. It can be shown, by an argument analogous to that of Lemma 7.11, that k_b is equal to $\mu(b) U_b(\mu_{|b}, \nu)$.   ⊣

Theorem 7.17   *Let $(Qx/W)\varphi$ be an IF formula in which x is public. Let $\mathbb{M}$ be a suitable structure and s an assignment in $\mathbb{M}$ whose domain contains $\mathrm{Free}((Qx/W)\varphi)$. Let $\Gamma_b = \Gamma(\mathbb{M}, s(x/b), \varphi)$ for each $b \in M$. Then*

$$V\Big(\Gamma(\mathbb{M}, s, (\exists x/W)\varphi)\Big) = \max\{V(\Gamma_b) \colon b \in M\},$$

$$V\Big(\Gamma(\mathbb{M}, s, (\forall x/W)\varphi)\Big) = \min\{V(\Gamma_b) \colon b \in M\}.$$

Proof   The proof is analogous to the proof of Theorem 7.12.   ⊣

We can use Theorem 7.17 to strip off quantifiers that quantify public variables.

Example 7.18   Consider the IF sentence φ,

$$\exists u \forall w (u \neq w \lor \varphi_{\mathrm{MP}}),$$

where φ_{MP} is the Matching Pennies sentence $\exists x (\forall y/\{x\}) x = y$. Let $\mathbb{M}$ be a finite structure with n elements. Since u is public in φ and w is public in $\forall w (u \neq w \lor \varphi_{\mathrm{MP}})$ it follows from Theorem 7.17 that

$$V\big(\Gamma(\varphi, \mathbb{M})\big) = \max_a \min_b V(\Gamma'),$$

where

$$\Gamma' = \Gamma\Big(\mathbb{M}, \{(u, a), (w, b)\}, u \neq w \lor \varphi_{\mathrm{MP}}\Big).$$

By the monotonicity rule for disjunction (Theorem 7.12), $V(\Gamma')$ equals

$$\max\Big[\Gamma\big(\mathbb{M}, \{(u, a), (w, b)\}, u \neq w\big), \Gamma\big(\mathbb{M}, \{(u, a), (w, b)\}, \varphi_{\mathrm{MP}}\big)\Big].$$

Hence, $V(\Gamma')$ is 1 if $a \neq b$ and $1/n$ otherwise. For every $a \in M$, Abelard can minimize $V(\Gamma')$ by choosing a for $\forall w$ as well. That is, $\min_b V(\Gamma')$ is $1/n$ for every $a \in M$. Consequently, $\max_a \min_b V(\Gamma')$ is $1/n$.   ⊣

7.2.6 Monotonicity rule for negation

Finally we will study the impact of negation on the value of IF formulas. We will show that Eloise can do no better in the game of φ than Abelard in the game of $\neg\varphi$. This principle is illustrated by the Matching Pennies sentence φ_{MP}. Its negation $\neg\varphi_{\text{MP}}$ is given by $\exists x(\forall y/\{x\})x \neq y$ and has the same value behavior as the Inverted Matching Pennies sentence φ_{IMP}. It follows from Examples 7.2 and 7.3 that the values of $\Gamma(\mathbb{M}, \varphi_{\text{MP}})$ and $\Gamma(\mathbb{M}, \varphi_{\text{IMP}})$ add up to 1.

First we prove an easy game-theoretic result.

Lemma 7.19 *Let* $\Gamma = \big(\{\text{I}, \text{II}\}, \{S_{\text{I}}, S_{\text{II}}\}, u\big)$ *be a two-person game and* $\Gamma' = \big(\{\text{I}, \text{II}\}, \{S'_{\text{I}}, S'_{\text{II}}\}, u'\big)$ *the two-person game obtained from* Γ *in the following way:* $S'_{\text{I}} = S_{\text{II}}$, $S'_{\text{II}} = S_{\text{I}}$ *and* $u(\sigma, \tau) = 1 - u'(\tau, \sigma)$ *for every* $\sigma \in S_{\text{I}}$ *and* $\tau \in S_{\text{II}}$. *Then* $V(\Gamma) = 1 - V(\Gamma')$.

Proof Let U and U' be the expected utility functions of Γ and Γ', respectively. Then

$$U(\mu, \nu) = \sum_{\sigma \in S_{\text{I}}} \sum_{\tau \in S_{\text{II}}} \mu(\sigma)\nu(\tau)\big(1 - u'(\tau, \sigma)\big) = 1 - U'(\nu, \mu).$$

Let (μ, ν) be an equilibrium in Γ. For every $\mu' \in \Delta(S_{\text{I}}) = \Delta(S'_{\text{II}})$, $U(\mu, \nu) \geq U(\mu', \nu)$ if and only if $1 - U'(\nu, \mu) \geq 1 - U'(\nu, \mu')$. Hence for every μ', $U'(\nu, \mu) \leq U'(\nu, \mu')$. In a similar fashion we can derive that for every $\nu' \in \Delta(S_{\text{II}}) = \Delta(S'_{\text{I}})$, $U'(\nu, \mu) \geq U'(\nu', \mu)$. It follows that (ν, μ) is an equilibrium in Γ'. $\dashv$

The following result shows that the values of φ and $\neg\varphi$ add up to 1 as claimed above.

Theorem 7.20 *Let* φ *be an IF formula,* $\mathbb{M}$ *a suitable structure, and* s *an assignment in* $\mathbb{M}$ *whose domain contains* Free(φ). *Then*

$$V\big(\Gamma(\mathbb{M}, s, \neg\varphi)\big) = 1 - V\big(\Gamma(\mathbb{M}, s, \varphi)\big).$$

Proof Every choice point in $G(\mathbb{M}, s, \varphi)$ belonging to p belongs to $\overline{p}$ in $G(\mathbb{M}, s, \neg\varphi)$ and *vice versa*. It follows that in the strategic IF game $\Gamma' = \Gamma(\mathbb{M}, s, \neg\varphi)$, Abelard (Eloise) controls the strategies that belong to Eloise (Abelard) in the strategic game $\Gamma = \Gamma(\mathbb{M}, s, \varphi)$.

Let σ be a strategy of Eloise in Γ and let τ be a strategy of Abelard in Γ. Let $h = (s, \varphi, \dots)$ be the terminal history in which Eloise follows σ and Abelard follows τ. The utility $u(\sigma, \tau)$ is determined by the assignment induced by h and the literal ψ for which $h \in H_\psi$, as follows: if $\mathbb{M}, s_h \models \psi$ then $u(\sigma, \tau) = 1$, and otherwise $u(\sigma, \tau) = 0$.

The strategy σ is a strategy of Abelard in Γ' and τ is a strategy of Eloise in the same game. Playing σ against τ yields the terminal history $h' = (s, \neg\varphi, \ldots) \in H_{\neg\psi}$. The assignments induced by h and h' coincide. Hence, if $\mathbb{M}, s_{h'} \models \neg\psi$ then $u'(\tau, \sigma) = 1$, and otherwise $u'(\tau, \sigma) = 0$. We conclude that $u(\sigma, \tau) = 1 - u'(\tau, \sigma)$, and the claim follows from Lemma 7.19. $\dashv$

7.3 Behavioral strategies and compositional semantics

The monotonicity rules in the previous section can be seen as a first attempt to find a compositional semantics for IF logic that is equivalent to equilibrium semantics. When evaluating IF formulas relative to single assignments under equilibrium semantics, we face the same difficulty we encountered in Chapter 4, namely that a single assignment is insufficient to encode the state of the semantic game. In fact, now the problem is even worse. Before (when the players were only allowed to use pure strategies) it was enough for the players to keep track of which assignments were possible. Now that they may use mixed strategies, the players must assign a probability to each assignment, as well.

Calculating how the probability of each assignment changes as the game progresses is nontrivial. The problem becomes more tractable if we assume the players follow behavioral strategies rather than mixed strategies. Whereas a mixed strategy randomizes a player's choice of pure strategy, a *behavioral strategy* randomizes the player's actions at each decision point. A *behavioral equilibrium* is defined similarly to a mixed-strategy equilibrium, i.e., no player can improve his or her expected utility by switching to a different behavioral strategy.

Independently of [54], Galliani [22] developed a probabilistic approach to IF logic based on behavioral strategies by assuming that when two histories are indistinguishable to a player, the probability distribution over his or her actions must be the same. He then gives a compositional semantics equivalent to the behavioral analogue of equilibrium semantics. He and the first author later generalized this semantics into a compositional semantics that is equivalent to equilibrium semantics, called *lottery semantics* [23].

Every behavioral strategy corresponds to a mixed strategy, but not conversely. The converse is true, however, for games with perfect recall.

This follows from a game-theoretic result known as Kuhn's theorem [35, 36]. Thus Galliani's semantics coincides with equilibrium semantics if we restrict to IF formulas for which both Eloise and Abelard have perfect recall.

7.4 Elimination of strategies

It is often clear to us that one strategy is inferior to another and that it is not in its owner's best interest to play it. We proceed to introduce the notion of "weak dominance" that formalizes this sense of a strategy being inferior to another.

Definition 7.21 Let Γ be a two-person strategic game. For $\sigma, \sigma' \in S_I$ we say that σ *weakly dominates* σ', or (equivalently) σ' is *weakly dominated* by σ if the following conditions are satisfied:

- For every $\tau \in S_{II}$, $u_I(\sigma, \tau) \geq u_I(\sigma', \tau)$.
- For some $\tau \in S_{II}$, $u_I(\sigma, \tau) > u_I(\sigma', \tau)$.

A similar notion is defined for player II. A strategy is *weakly dominated* in Γ if its owner has a strategy in Γ that weakly dominates it. ⊣

Suppose Eloise's strategy σ weakly dominates[1] the strategy σ' in the strategic IF game Γ. Then, by definition, for every strategy τ of Abelard, $u(\sigma, \tau) \geq u(\sigma', \tau)$. Since in strategic IF games either player receives 0 or 1, this means that for every τ, if $u(\sigma', \tau) = 1$ then $u(\sigma, \tau) = 1$. That is, if the weakly dominated σ' wins against τ, then so does σ. Furthermore, by definition, there is a strategy τ for which $u(\sigma, \tau) > u(\sigma', \tau)$, i.e, there is a strategy against which σ wins, but σ' loses.

Our interest is in finding the value of strategic IF games, that is, in finding equilibria. To this end, it is useful to eliminate certain strategies so that we obtain a smaller, less complex game that has the same value. The following result enables us to eliminate weakly dominated strategies.

Proposition 7.22 *Let Γ be a finite, two-player, strictly competitive strategic game. Then Γ has an equilibrium (μ_I, μ_{II}) such that for each*

---

[1] In the literature, weak dominance is contrasted to strong dominance. A strategy σ of player p *strongly dominates* another strategy σ' if the following condition is met:

- For every $\tau \in S_{\overline{p}}$, $u_p(\sigma, \tau) > u_p(\sigma', \tau)$.

player p, none of the strategies in the support of μ_p is weakly dominated in Γ.

Proof Suppose (μ, ν) is an equilibrium in Γ. Suppose ρ and ρ' are two strategies in the support of μ, such that ρ weakly dominates ρ'. Consider the mixed strategy μ^* in Γ that is defined for $\sigma \in S_{\mathrm{I}}$ by

$$\mu^*(\sigma) = \begin{cases} \mu(\rho) + \mu(\rho') & \text{if } \sigma \text{ is } \rho \\ 0 & \text{if } \sigma \text{ is } \rho' \\ \mu(\sigma) & \text{otherwise.} \end{cases}$$

So μ^* is the mixed strategy that is exactly like μ except that it plays ρ whenever μ would play ρ'. The support of μ^* is that of μ minus ρ'.

By the definition of equilibrium, $U_{\mathrm{II}}(\mu, \nu) \geq U_{\mathrm{II}}(\mu, \nu')$ for every $\nu' \in \Delta(S_{\mathrm{II}})$. Since Γ is strictly competitive,

$$U_{\mathrm{I}}(\mu, \nu) \leq U_{\mathrm{I}}(\mu, \nu'). \tag{7.3}$$

Now we will show that

$$U_{\mathrm{I}}(\mu, \nu') \leq U_{\mathrm{I}}(\mu^*, \nu'), \tag{7.4}$$

which coincides with

$$\sum_{\sigma \in S_{\mathrm{I}}} \sum_{\tau \in S_{\mathrm{II}}} \mu(\sigma)\nu'(\tau)u_{\mathrm{I}}(\sigma, \tau) \leq \sum_{\sigma \in S_{\mathrm{I}}} \sum_{\tau \in S_{\mathrm{II}}} \mu^*(\sigma)\nu'(\tau)u_{\mathrm{I}}(\sigma, \tau). \tag{7.5}$$

Since μ and μ^* agree on every strategy except ρ and ρ', it follows from Equation (7.5) that

$$\left[\sum_{\tau \in S_{\mathrm{II}}} \mu(\rho)\nu'(\tau)u_{\mathrm{I}}(\rho, \tau) \right] + \left[\sum_{\tau \in S_{\mathrm{II}}} \mu(\rho')\nu'(\tau)u_{\mathrm{I}}(\rho', \tau) \right] \tag{7.6}$$

is less than or equal to

$$\left[\sum_{\tau \in S_{\mathrm{II}}} \mu^*(\rho)\nu'(\tau)u_{\mathrm{I}}(\rho, \tau) \right] + \left[\sum_{\tau \in S_{\mathrm{II}}} \mu^*(\rho')\nu'(\tau)u_{\mathrm{I}}(\rho', \tau) \right]. \tag{7.7}$$

By definition, $\mu^*(\rho) = \mu(\rho) + \mu(\rho')$ and $\mu^*(\rho') = 0$. Thus, modulo some rewriting, (7.7) is equal to

$$\left[\sum_{\tau \in S_{\mathrm{II}}} \mu(\rho)\nu'(\tau)u_{\mathrm{I}}(\rho, \tau) \right] + \left[\sum_{\tau \in S_{\mathrm{II}}} \mu(\rho')\nu'(\tau)u_{\mathrm{I}}(\rho, \tau) \right]. \tag{7.8}$$

Since ρ weakly dominates ρ', $u_{\mathrm{I}}(\rho', \tau) \leq u_{\mathrm{I}}(\rho, \tau)$ for every $\tau \in S_{\mathrm{II}}$. Hence, (7.6) is less than or equal to (7.8), and Equation (7.4) follows.

If we substitute ν for ν' in Equation (7.4), we get $U_{\text{I}}(\mu, \nu) \leq U_{\text{I}}(\mu^*, \nu)$. Since (μ, ν) is an equilibrium, we also have $U_{\text{I}}(\mu, \nu) \geq U_{\text{I}}(\mu^*, \nu)$. Hence $U_{\text{I}}(\mu, \nu) = U_{\text{I}}(\mu^*, \nu)$. Connecting this equivalence with Equations (7.3) and (7.4) yields

$$U_{\text{I}}(\mu^*, \nu) = U_{\text{I}}(\mu, \nu) \leq U_{\text{I}}(\mu, \nu') \leq U_{\text{I}}(\mu^*, \nu').$$

Since Γ is strictly competitive, we arrive at $U_{\text{II}}(\mu^*, \nu) \geq U_{\text{II}}(\mu^*, \nu')$. $\dashv$

A result similar to Proposition 7.22 can be proved for finite, non-strictly competitive games, but the proof is more complicated, cf. [44, p. 122].

Let

$$\Gamma = \big(N, \{S_p : p \in N\}, \{u_p : p \in N\}\big)$$

be a finite, two-player, strictly competitive game in which $S_{\text{I}}^* \subseteq S_{\text{I}}$ is the set of player I's strategies that are not weakly dominated in Γ. By Proposition 7.22, if (μ, ν) is an equilibrium in Γ, then there is a mixed strategy $\mu^* \in \Delta(S_{\text{I}}^*)$ such that (μ^*, ν) is an equilibrium in Γ. Any two equilibria in Γ have the same value (Proposition 2.15). Consequently, (μ, ν) and (μ^*, ν) have the same value. The same reasoning applies to player II's strategies in $S_{\text{II}}^* \subseteq S_{\text{II}}$ that are not weakly dominated in Γ.

It follows that the smaller game

$$\Gamma^* = \big(N, \{S_p^* : p \in N\}, \{u_p^* : p \in N\}\big),$$

where u_p^* is the restriction of u_p to the pairs of strategies in Γ^*, has the same value as Γ. We can iterate the elimination process to obtain a series $\Gamma^*, \Gamma^{**}, \Gamma^{***}, \ldots$ of games of decreasing size that all have the same value.

Note that if Γ is a strategic IF game and Γ^* is a subgame of Γ that is the result of removing weakly dominated strategies, then Γ^* need not be a strategic IF game. That is, the class of strategic IF games is not closed under removing weakly dominated strategies.

We present another result that enables us to remove strategies that are equivalent in terms of payoff. Let Γ be a two-player strategic game and let $\sigma, \sigma' \in S_{\text{I}}$. We say σ and σ' are *payoff equivalent* if for every strategy τ belonging to player II, $u_{\text{I}}(\sigma, \tau) = u_{\text{I}}(\sigma', \tau)$. A similar notion is defined for player II.

Proposition 7.23 *Let Γ be a strategic game. Then Γ has an equilibrium $(\mu_{\text{I}}, \mu_{\text{II}})$ such that for each player, there are no two strategies in the support of μ_p that are payoff equivalent.*

Proof Analogous to the proof of Proposition 7.22. As a matter of fact we can prove that $U_p(\mu_I, \mu'_{II}) = U_p(\mu_I^*, \mu'_{II})$, which is a stronger statement than what we need for Equation (7.4). ⊣

Example 7.24 Consider the two-player, zero-sum game Γ_0 in Figure 7.4(a). In this game, player II's strategy τ_4 is weakly dominated by τ_2. Hence, the value of Γ_0 coincides with that of Γ_1 in Figure 7.4(b). In Γ_1, σ_4 is weakly dominated by σ_1. (Note that σ_4 is not weakly dominated by σ_1 in Γ_0.) Therefore we can eliminate σ_4 and obtain the game Γ_2, which has the same value as Γ_0 and Γ_1. The strategy τ_1 is weakly dominated in Γ_2 by τ_3. Removing it yields Γ_3, as shown in Figure 7.4(d). In Γ_3, σ_3 is weakly dominated by σ_1 and σ_2. If we remove σ_3, we obtain Γ_4, a variant of the Matching Pennies game, as shown in Figure 7.4(e), which cannot be further reduced by eliminating weakly dominated strategies. It follows that the value of Γ_0 is $1/2$.

In a strategic game there may be many strategies that can be eliminated. The order in which we eliminate them has an effect on the game in which we end up. For instance, in Γ_2, σ_1 weakly dominates σ_3. Eliminating σ_3 yields Γ'_3, as shown in Figure 7.4(f). In Γ'_3, the strategies τ_1 and τ_3 are payoff equivalent. Eliminating the latter yields Γ'_4, which is again a variant of the Matching Pennies game, as shown in Figure 7.4(g). ⊣

Observe that in strategic IF games, σ is weakly dominated by or payoff equivalent to σ' if and only if for each strategy τ, $u(\sigma, \tau) \leq u(\sigma', \tau)$. Similarly, τ is weakly dominated by or payoff equivalent to τ' if and only if for each strategy σ, $u(\sigma, \tau) \geq u(\sigma, \tau')$.

If σ is a winning strategy of player p in the extensive game $G(\mathbb{M}, s, \varphi)$ then it weakly dominates all strategies of p in $\Gamma(\mathbb{M}, s, \varphi)$ that are not payoff equivalent to σ. Thus we can eliminate all strategies of p other than the winning strategy σ without harming her expected utility. In the reduced game, her opponent $\bar{p}$ still owns all the strategies he owned in the original game. However, in the reduced game, they are all payoff equivalent to each other as they all lose against σ. Therefore we can eliminate all but one arbitrary strategy of player $\bar{p}$. In the resulting game both players have one choice.

Every semantic game associated with a first-order formula is determined and its strategic games can therefore be reduced to a "1×1 game." By abuse of terminology, we will refer to a subformula of φ with empty slash sets as a first-order subformula of φ. If ψ is a first-order formula that appears as a subformula in the IF formula φ, then we can

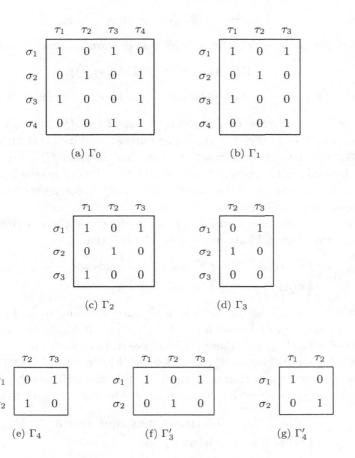

Figure 7.4 A series of strategic games in which Γ_i is obtained from Γ_{i-1} by eliminating weakly dominated strategies. The process is non-deterministic since both Γ_3 and Γ_3' can be obtained in this manner from Γ_2.

reduce every strategic IF game of φ to a smaller game in which for every player p, all of p's strategies yield the same actions for p's choice points in ψ. This amounts to treating ψ as an atomic formula.

Proposition 7.25 *Let φ be an IF_L formula and ψ a first-order sub-formula in φ such that $\text{Free}(\psi) = \{x_1, \ldots, x_m\}$. Let $R \notin L$ be an m-ary relation symbol, and $\mathbb{M}$ an $L \cup \{R\}$-structure such that for every assignment s in $\mathbb{M}$ with domain containing $\text{Free}(\psi)$,*

$$\mathbb{M}, s \models \psi \quad \text{iff} \quad \mathbb{M}, s \models R(x_1, \ldots, x_m).$$

Then for every assignment s in $\mathbb{M}$ with domain containing Free(φ),

$$V\big(\Gamma(\mathbb{M}, s, \varphi)\big) = V\big(\Gamma(\mathbb{M}, s, \varphi')\big),$$

where φ' is the result of replacing ψ in φ by $R(x_1, \ldots, x_m)$.

Proof In the game $G = G(\mathbb{M}, s, \varphi)$, let $H_\psi = \{h_1, \ldots, h_n\}$ and H'_ψ the closure of histories in H_ψ under extensions. Since ψ is first-order it follows from the Gale-Stewart theorem that the game $G_i = G(\mathbb{M}, s_{h_i}, \psi)$ is determined for every $1 \leq i \leq n$. Without loss of generality, assume that Eloise has winning strategies $\sigma_1^*, \ldots, \sigma_j^*$ in the games $G_1, \ldots, G_j$, respectively, for some $0 \leq j \leq n$.

For every strategy σ in Γ let σ^* be the function whose arguments are the histories h of Eloise in the game G such that

$$\sigma^*(h) = \begin{cases} \sigma_i^*(s_h, \psi, \ldots) & \text{if } h \text{ equals or extends some } h_i \in \{h_1, \ldots, h_j\}, \\ \sigma(h) & \text{otherwise,} \end{cases}$$

where $(s_h, \psi, \ldots)$ is the history in G_i corresponding to h. To see that σ^* is a strategy of Eloise in $\Gamma = \Gamma(\mathbb{M}, s, \varphi)$ let $h \approx_W h'$ be two distinct histories in $H_{(\exists x/W)\chi}$. Since h and h' are distinct and ψ is first-order, h and h' do not belong to H'_ψ. Therefore, $\sigma^*(h) = \sigma(h)$ and $\sigma^*(h') = \sigma(h')$. The strategy σ is certainly W-uniform. Whence, $\sigma(h) = \sigma(h')$.

Similarly, we let τ^* be the strategy of Abelard in Γ such that

$$\tau^*(h) = \begin{cases} \tau_i^*(s_h, \psi, \ldots) & \text{if } h \text{ equals or extends some } h_i \in \{h_{j+1}, \ldots, h_n\}, \\ \tau(h) & \text{otherwise,} \end{cases}$$

where $(s_h, \psi, \ldots)$ as above.

To see that σ is payoff equivalent to or weakly dominated by σ^*, fix any strategy τ of Abelard. Let h be the history that follows σ and τ. If h extends a history in $\{h_1, \ldots, h_j\}$, playing σ^* against τ yields a history that is winning for Eloise. Otherwise, if h does not extend a history in $\{h_1, \ldots, h_j\}$, playing σ^* against τ yields h as well. Therefore, $u(\sigma^*, \tau) \geq u(\sigma, \tau)$ for every τ.

In the same way we can see that any strategy τ of Abelard is payoff equivalent to or weakly dominated by τ^*.

Let $S^* = \{\sigma^* : \sigma \in S\}$ and $T^* = \{\tau^* : \tau \in T\}$. By Propositions 7.22 and 7.23, the game $\Gamma^* = (\{\exists, \forall\}, \{S^*, T^*\}, u^*)$ has the same value as Γ, where u^* is the function defined on $S^* \times T^*$ such that $u^*(\sigma^*, \tau^*) = u(\sigma^*, \tau^*)$.

We will show that Γ^* and $\Gamma' = \Gamma(\mathbb{M}, s, \varphi')$ are the same game modulo renaming of strategies. To this end, observe that every history in G that is not in H'_ψ is a history in $G' = G(\mathbb{M}, s, \varphi')$ and *vice versa*. Further, we observe that every two strategies σ_k and σ_ℓ of Eloise in Γ that make the same action on every history $h \notin H'_\psi$ are mapped to the same strategy $\sigma_k^* = \sigma_\ell^*$. So for any two strategies of Eloise in Γ^* there exists a history $h \notin H'_\psi$ such that $\sigma_k^*(h) \neq \sigma_\ell^*(h)$. The same applies to Abelard's strategies.

For every strategy σ' of Eloise in Γ' let $\beta(\sigma')$ be the strategy in Γ^* such that

$$\beta(\sigma')(h) = \begin{cases} \sigma'(h) & \text{if } h \notin H'_\psi, \\ \sigma^*(h) & \text{otherwise,} \end{cases}$$

for any σ^* from Γ^*. The function β is injective, since for any two strategies $\sigma_1' \neq \sigma_2'$ of Eloise in Γ', clearly $\beta(\sigma_1') \neq \beta(\sigma_2')$. It is also surjective, since for every strategy σ^* of Eloise in Γ^* it is the case that $\beta(\sigma') = \sigma^*$, where σ' is the strategy in Γ' such that for every history $h \notin H'_\psi$ of Eloise, $\sigma^*(h) = \sigma'(h)$. Hence β is a bijection between the strategies in S^* and the strategies of Eloise in Γ'. Similarly we can show that there is a bijection β between the strategies in T^* and the strategies of Abelard in Γ'.

Consider any pair of strategies (σ^*, τ^*) from $S^* \times T^*$. Let h be the history that is the result of playing σ^* against τ^*. Suppose $h \notin H'_\psi$. Then, playing $\beta(\sigma^*)$ against $\beta(\tau^*)$ also yields h. Hence, $u^*(\sigma^*, \tau^*) = u'\big(\beta(\sigma^*), \beta(\tau^*)\big)$, where u' is the utility function of Γ'.

Suppose $h \in H'_\psi$, in particular suppose h equals or extends the history $h_i \in H_\psi$. Assume that h is winning for Eloise. Then Eloise has a winning strategy in G_i and therefore $\mathbb{M}, s_{h_i} \models \psi$. By hypothesis,

$$\mathbb{M}, s_{h_i} \models R(x_1, \ldots, x_m).$$

The history h' that is the result of playing $\beta(\sigma^*)$ against $\beta(\tau^*)$ sits in $H_{R(x_1, \ldots, x_m)}$ and furthermore $s_{h_i} = s_{h'}$. We conclude that

$$\mathbb{M}, s_{h'} \models R(x_1, \ldots, x_m).$$

The case for Abelard is similar and therefore $u^*(\sigma^*, \tau^*)$ is equal to $u'\big(\beta(\sigma^*), \beta(\tau^*)\big)$.

Modulo renaming strategies, Γ^* and Γ' are identical. In particular, Γ^* and Γ' have the same value. Since Γ and Γ^* have the same value, also Γ and Γ' have the same value.   ⊣

Example 7.26 Consider the following IF sentence φ,

$$\exists x \big(\forall y/\{x\}\big)(x = c \vee x = y).$$

Let $\mathbb{M}$ be a finite structure with n elements. In the strategic game $\Gamma = \Gamma(\mathbb{M}, \varphi)$, Eloise has n choice functions associated with $\exists x$ and no less than 2^{n^2} choice functions associated with the disjunction $(x = c \vee x = y)$. So in total she has $n2^{n^2}$ strategies in Γ. Let R be a fresh binary relation symbol. Let $\mathbb{M}^*$ be the expansion of $\mathbb{M}$ such that

$$\mathbb{M}^*, s \models x = c \vee x = y \quad \text{iff} \quad \mathbb{M}^*, s \models R(x, y),$$

for every assignment s in $\mathbb{M}$ with domain containing $\{x, y\}$. Let φ' be the IF sentence

$$\exists x \big(\forall y/\{x\}\big) R(x, y).$$

In the game $\Gamma' = \Gamma(\mathbb{M}^*, \varphi')$, Eloise has only n strategies. By Proposition 7.25, the value of $\Gamma(\mathbb{M}^*, \varphi)$ is equal to the value of $\Gamma(\mathbb{M}^*, \varphi')$.   ⊣

The following example illustrates how we can use iterated elimination of payoff equivalent and weakly dominated strategies to compute the value of a complex strategic IF game.

Example 7.27 Write θ_∞ for the IF sentence $\exists w \theta'_\infty$, where θ'_∞ is

$$\forall x \exists y \big(\exists z/\{x\}\big)(z = x \wedge y \neq w).$$

Eloise has a winning strategy in the game $G = G(\mathbb{M}, s, \theta_\infty)$ if $\mathbb{M}$ is (Dedekind) infinite (recall the sentence φ_∞ in Example 4.14). It can be checked that Abelard has a winning strategy in G if $\mathbb{M}$ contains one element; G is undetermined in all other cases. Suppose that $\mathbb{M}$ is finite and contains more than one element.

Since w is public in θ_∞, it follows from the monotonicity rule for quantifiers (Theorem 7.17) that $V(\Gamma) = \max\{\Gamma_d \colon d \in M\}$, where $\Gamma_d = \Gamma(\mathbb{M}, s, \theta'_\infty)$ and $s = \{(w, d)\}$. Fix $d \in M$ and consider the game Γ_d.

Let θ^*_∞ be the result of replacing the subformula $(z = x \wedge y \neq w)$ by the atomic formula $R(x, y, z, w)$, where R is a fresh 4-ary relation symbol. Since the formula $(z = x \wedge y \neq w)$ is first-order, it follows from Proposition 7.25 that the value of Γ_d is equal to the value of $\Gamma^*_d = \Gamma(\mathbb{M}^*, s, \theta^*_\infty)$ where $\mathbb{M}^*$ is the expansion of $\mathbb{M}$ that interprets R as in the hypothesis of Proposition 7.25.

Let S be the set of strategies of Eloise in Γ_d^* and T the set of strategies of Abelard. We index each of Abelard's strategies $\tau_a \in T$ with the object Abelard chooses, that is,

$$\tau_a(s, \theta_\infty^*) = (x, a).$$

We associate with every strategy $\sigma \in S$ of Eloise the pair of choice functions (f, g), where $f \colon M \to M$ is the choice function associated with $(\exists y / \{w\})$ such that

$$\sigma(s, \theta_\infty^*, (x, a)) = (y, f(a)),$$

and $g \colon M \to M$ is the choice function associated with $(\exists z / \{w, x\})$ such that

$$\sigma(s, \theta_\infty^*, (x, a), (y, b)) = (z, g(b)).$$

For $A \subseteq M$, let $S_A \subseteq S$ be the set of strategies (f, g) such that

$$g(f(a)) = a \text{ and } f(a) \neq d \quad \text{iff} \quad a \notin A.$$

Every $\sigma \in S_A$ wins against $\tau_a \in T$ if and only if $a \notin A$. Since Eloise does not have a winning strategy, S_A is empty for $A = \varnothing$.

Let $A \subseteq M$ and $\sigma, \sigma' \in S_A$. To see that σ and σ' are payoff equivalent, consider an arbitrary strategy $\tau_a \in T$. By the construction of S_A, $u(\sigma, \tau) = 1$ if and only if $a \notin A$ and by a similar argument: $u(\sigma', \tau) = 1$ if and only if $a \notin A$. Hence, $u(\sigma, \tau) = u(\sigma', \tau)$ for all $\tau \in T$.

Let S' be a set that contains precisely one strategy $\sigma = (f, g)$ from $S_{\{a\}}$ for each $a \in M$. It is not hard to see that $S_{\{a\}}$ is nonempty as it contains the pair (f, g), where $f = g$ are the functions on M such that

$$f(z) = \begin{cases} a & \text{if } z = d, \\ d & \text{if } z = a, \\ z & \text{otherwise.} \end{cases}$$

Figure 7.5 provides an illustration of what f is like on a five-element structure. Since f is an involution, $g(f(z)) = z$ for every $z \in M$. In addition, $f(z) = d$ if and only if $z = a$. Hence, $(f, g) \in S_{\{a\}}$.

Let B be a nonempty subset of M. Fix any $\sigma \in S_B$. For any $\tau_c \in T$, suppose $u(\sigma, \tau) = 1$, that is, $c \notin B$. For any singleton $\{a\} \subseteq B$ we also have that $c \notin \{a\}$. Hence, for the strategy σ' in $S' \cap S_{\{a\}}$, $u(\sigma', \tau) = 1$. It follows that every strategy σ in S is payoff equivalent to or weakly dominated by a strategy in S'.

By Propositions 7.22 and 7.23, it follows that there exists a $\mu \in \Delta(S')$ such that (μ, ν) is an equilibrium in Γ_d^*.

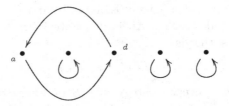

Figure 7.5 The function f on a five-element structure. It is an involution that maps a to d and the other objects to themselves.

Let us focus on the payoff matrix defined by $S' \times T$. We already saw that $T = \{b\colon b \in M\}$. Similarly, S' can be written as $\{\sigma_a\colon a \in M\}$ where σ_a is the strategy from $S_{\{a\}}$. Consider the strategies σ_a and τ_b. Eloise wins if and only if $a \neq b$. It follows that the payoff matrix of $S' \times T$ is equal to the payoff matrix of the Inverted Matching Pennies game depicted in Figure 7.1(b), modulo renaming of strategies. Hence, the value of Γ_d is $(n-1)/n$. Since d was chosen arbitrarily, the value of Γ is $(n-1)/n$ as well.

Interestingly, on finite structures θ_∞ has the same value behavior as the simpler

$$\varphi_{\mathrm{IMP}} = \forall x (\exists z/\{x\}) x \neq z.$$

Apparently, the signaling quantifier $\exists y$ and the conjunct $y \neq w$ only come into action on infinite structures.   ⊣

The value of $\Gamma(\mathbb{M}, \theta_\infty)$ increases as n goes to infinity. It is tempting to take the value of the game on finite models as an approximation of the property expressed by θ_∞. This may be true in this case but it does not hold in general. For instance, [54] contains as an example the IF sentence φ_{even},

$$\forall x \forall y (\exists u/\{y\}) (\exists v/\{x\}) (v = x \wedge u = y \wedge u \neq x \wedge x = y \to u = v),$$

which is true on finite $\mathbb{M}$ if and only if M has even cardinality. It was shown that if $\mathbb{M}$ has odd cardinality n, then the value of $\Gamma(\mathbb{M}, \varphi_{\mathrm{even}})$ is $1 - (1/2n)$. The value of the strategic game $\Gamma(\mathbb{M}, \varphi_{\mathrm{even}})$ increases as n grows, for odd n. However, in the case of φ_{even} it does not make much sense to say that a structure with nine objects is a closer approximation of the property of having odd cardinality than a structure with seven objects.

7.5 Expressing the rationals

In this section we will study the values that the strategic IF games can realize. We show that they realize precisely the rationals in the interval $[0, 1]$.

First we recall a result about the values of general zero-sum, two-player games, see [48, p. 739].

Theorem 7.28 *Let Γ be a zero-sum, two-player game whose utility function has the rationals as range. Then the value of Γ is rational.*

We have already seen an IF sentence that can realize all rationals. The IF sentence

$$\exists x (\forall y / \{x\}) P((x + y) \bmod n),$$

which defines the Rescher quantifier (see Example 7.7), has value m/n, where m the number of P objects and n the size of the universe at hand. Hence, for every rational m/n there is a strategic IF game with m/n as its value.

In the remainder of this section we show a stronger version of this result. We show that for every rational q in the interval $[0, 1]$ there is an IF sentence in the empty vocabulary that has value q on every structure with two or more elements.

Theorem 7.29 (Sevenster and Sandu [54, Theorem 19]) *Let $0 \leq m < n$ be integers and $q = m/n$. There exists an IF sentence that has value q on every structure with at least two objects.*

Proof The idea of the proof is as follows. We let one player pick m objects from a set of n objects. Then we let another player pick one object from the same set of n objects. The probability that the second player picked an object that was among the first player's m objects is obviously m/n. We will define a game in the language of IF logic that enforces this behavior. The value of this game will be m/n and the result follows.

- *Game* — Let M be a set of at least n objects and C a subset of M with precisely n objects. We describe a two-step game:

S1: Abelard picks m objects $b_1, \ldots, b_m$ from M;

S2: Eloise picks one object c from M not knowing the objects chosen in S1.

    Eloise gets payoff 1 if and only if at least one of the following conditions is met for some distinct $1 \leq i, j \leq m$:

(1) Abelard has chosen two equal objects: $b_i = b_j$;

(2) Abelard has chosen outside C: $b_i \notin C$;

(3) Eloise guesses one of Abelard's objects: $c = b_i$.

Conditions (1) and (2) force Abelard to chose m distinct objects from C. Let G denote the game thus described. There is no need to hardwire in the winning conditions that Eloise must choose c from C. Any strategy of Eloise that chooses c from outside C is weakly dominated by every strategy that chooses an object from C.

- *Value* — Let μ be Eloise's uniform mixed strategy with support $\{\sigma_a : a \in C\}$. Let $\mathcal{B}$ be the set of sets B for which $B \subseteq C$ and $|B| = m$. Let T_B be the set of strategies τ that pick the objects in B (in any order). All strategies in T_B are payoff equivalent. Every strategy τ' that is not in a set T_B violates condition (1) and/or (2) and will result in a loss for Abelard. Every such strategy τ' is weakly dominated by a strategy from T_B for any $B \in \mathcal{B}$. Pick one strategy τ_B from T_B for each $B \in \mathcal{B}$, and collect these strategies in T^*. Since every strategy in T is payoff equivalent to or weakly dominated by a strategy from T^*, the value of G is equal to the value of the game in which Abelard chooses from T^* instead of T.

  Let ν be Abelard's uniform mixed strategy with support T^*. We claim that (μ, ν) is an equilibrium in G with value m/n. First we observe that

  $$U(\mu, \nu) = \sum_{a \in C} \sum_{B \in \mathcal{B}} \mu(\sigma_a)\nu(\tau_B)u(\sigma_a, \tau_B) = \sum_{B \in \mathcal{B}_a} \nu(\tau_B),$$

  where $\mathcal{B}_a = \{B' \in \mathcal{B} : a \in B'\}$ for any object $a \in C$. The expression $\sum_{B \in \mathcal{B}_a} \nu(\tau_B)$ denotes the probability m/n that a randomly drawn set B of m objects contains a. The payoff matrix of the reduced game has the same diagonal band of 1s as Figure 7.2. It can thus we seen that (μ, ν) is an equilibrium.

- *Definition in* IF *logic* — Assume that $\mathbb{M}$ interprets the constant symbols $c_1, \ldots, c_n$ in such a way that $C = \{c_1^{\mathbb{M}}, \ldots, c_n^{\mathbb{M}}\}$.

  The following IF sentence defines G:

  $$\forall x_1 \ldots \forall x_m (\exists y/\{x_1, \ldots, x_m\})\beta_1 \vee \beta_2 \vee \beta_3,$$

where

$$\beta_1 \text{ is } \bigvee_{i\in\{1,\ldots,m\}} \bigvee_{j\in\{1,\ldots,m\}-\{i\}} x_i = x_j,$$

$$\beta_2 \text{ is } \bigvee_{i\in\{1,\ldots,m\}} \bigwedge_{j\in\{1,\ldots,n\}} x_i \neq c_j,$$

$$\beta_3 \text{ is } \bigvee_{i\in\{1,\ldots,m\}} x_i = y.$$

The formulas β_1–β_3 encode the respective winning conditions (1)–(3) from G in first-order logic. By Proposition 7.25 we can replace the first-order subformula $\beta_1 \vee \beta_2 \vee \beta_3$ by the atom $R(x_1,\ldots,x_m,y)$ to obtain an IF sentence ψ that has the same value on $\mathbb{M}^*$ as $\mathbb{M}$, where $\mathbb{M}^*$ is the appropriate expansion of $\mathbb{M}$. The game of ψ is the game described earlier.

The previous game assumes that M contains at least n objects and that we have n constants at our disposal, each interpreted by a different object. We can drop both assumptions by letting Eloise pick any set of n objects from which Abelard has to chose, and to let them draw bitstrings that encode their choice. In this way, we only require two dedicated objects (the *bits*), which can also be chosen by Eloise. It does not matter which two objects she choses as long as they are distinct.

- *Game* — Let $\ell = \lceil 2\log n \rceil$.

S0′: Eloise picks two distinct objects a_0 and a_1 from M;

S1′: Eloise picks ℓn objects $b_1^1,\ldots,b_1^\ell,\ldots,b_n^1,\ldots,b_n^\ell$ from M;

S2′: Abelard picks ℓm objects $c_1^1,\ldots,c_1^\ell,\ldots,c_m^1,\ldots,c_m^\ell$ from M;

S3′: Eloise picks ℓ objects $d^1,\ldots,d^\ell$ from M not knowing the objects chosen in S2′.

Eloise gets payoff 1 if and only if $a_0 \neq a_1$, $b_1^1,\ldots,b_n^\ell,d^1,\ldots,d^\ell \in \{a_0,a_1\}$ and at least one of the following conditions is met for some distinct $1 \leq i,j \leq m$:

(1) Abelard has chosen two equal bitstrings: $c_i^k = c_j^k$ for all $1 \leq k \leq \ell$;

(2) One of Abelard's bitstrings is not in $\{(b_1^1,\ldots,b_1^\ell),\ldots,(b_n^1,\ldots,b_n^\ell)\}$:

$$(c_i^1,\ldots,c_i^\ell) \notin \{(b_1^1,\ldots,b_1^\ell),\ldots,(b_n^1,\ldots,b_n^\ell)\};$$

(3) Eloise guesses one of Abelard's bitstrings: $c_i^k = d^k$ for all $1 \leq k \leq \ell$.

Let G' be the game described by S0′–S3′ and the latter winning conditions.

- *Value* — Compared to G, Eloise has an extra step in G', but there is no way in which she can benefit from it. The other steps correspond to the steps in G, modulo the coding trick by bitstrings. Hence the value of G' is m/n.
- *Definition in* IF *logic* — The following IF sentence defines G':

$$\exists z_1 \exists z_2 \exists w_1^1 \ldots \exists w_1^\ell \ldots \exists w_n^1 \ldots \exists w_n^\ell \forall x_1^1 \ldots \forall x_1^\ell \ldots \forall x_m^1 \ldots \forall x_m^\ell$$
$$(\exists y^1/X) \ldots (\exists y^\ell/X)\gamma,$$

where $X = \{x_1^1, \ldots, x_m^\ell\}$ and $\gamma = \big(z_1 \neq z_2 \wedge (\gamma_1 \vee \gamma_2 \vee \gamma_3)\big)$ with

$$\gamma_1 \text{ is } \bigvee_{i \in \{1,\ldots,m\}} \bigvee_{j \in \{1,\ldots,m\}-\{i\}} (x_i^1 = x_j^1 \wedge \ldots \wedge x_i^\ell = x_j^\ell)$$

$$\gamma_2 \text{ is } \bigvee_{i \in \{1,\ldots,m\}} \bigwedge_{j \in \{1,\ldots,n\}} (w_j^1 \neq x_i^1 \vee \ldots \vee w_j^\ell \neq x_i^\ell)$$

$$\gamma_3 \text{ is } \bigvee_{i \in \{1,\ldots,m\}} (x_i^1 = y^1 \wedge \ldots \wedge x_i^\ell = y^\ell).$$

Observe that the variables z_i are public in the subformula $\exists z_i \ldots$, and similarly for the variables w_j^h in the subformula $\exists w_j^h \ldots$. $\quad\dashv$

8
Further topics

In our final chapter we briefly discuss two topics we feel deserve to be mentioned in spite of space constraints that prevent us from giving them fuller treatment. In the first half of the chapter, we address the debate concerning the possibility of giving a compositional semantics for IF logic. In the second half, we investigate the effect of introducing imperfect information to modal logic.

8.1 Compositionality

The original game-theoretic semantics for IF logic assigned meanings only to IF sentences [30]. Thus IF logic was immune to a common complaint lodged against Tarski's semantics for first-order logic, namely that truth is defined in terms of satisfaction, rather than truth alone. However, it also meant that one was not able to analyze IF sentences by looking at the meanings of their subformulas. Furthermore, Hintikka famously claimed that there could be no compositional semantics for IF logic:

... there is no realistic hope of formulating compositional truth-conditions for [IF sentences], even though I have not given a strict impossibility proof to that effect. [28, pp. 110ff][1]

Hintikka's assertion inspired Hodges to develop his trump semantics, which gives meanings to all IF formulas [32, 33]. In Chapter 4, we defined two other semantics for IF formulas: game-theoretic semantics and Skolem semantics. In order to emphasize the similarities between IF

---

[1] See also [31, pp. 370–371].

logic and first-order logic, we introduced both semantics in terms of single assignments. However, in order to prove their equivalence with trump semantics, we later extended them to teams of assignments.

Is such an extension necessary? In other words, can one formulate a compositional semantics for IF logic that defines satisfaction (and dissatisfaction) using single assignments? Before we can attempt to answer such questions, we must specify the properties we desire such a semantics to have.

Definition 8.1 (Cameron and Hodges [11])   A *semantics* for IF logic is a function that assigns to every IF formula φ a *meaning* $|\varphi|_{\mathbb{M},U}$ in every suitable structure $\mathbb{M}$, relative to any finite set of variables U containing Free(φ). We will simply write $|\varphi|_U$ when the structure is clear from context, and omit U when it is empty. A semantics for IF logic is *adequate* if it has the following two properties:

(1) There is a distinguished value, called TRUE, such that for any IF sentence φ we have $|\varphi|_{\mathbb{M}} = $ TRUE if and only if $\mathbb{M} \models^+_{\mathrm{GTS}} \varphi$.

(2) For any IF sentence φ with a subformula ψ, if φ' is the result of replacing ψ by an IF formula ψ', $U = A_\varphi(\psi) = A_{\varphi'}(\psi')$ is the set of variables quantified superordinate to ψ in φ, and

$$|\psi|_{\mathbb{M},U} = |\psi'|_{\mathbb{M},U},$$

then $|\varphi|_{\mathbb{M}} = $ TRUE if and only if $|\varphi'|_{\mathbb{M}} = $ TRUE.   ⊣

Condition (1) says that an adequate semantics for IF logic agrees with game-theoretic semantics on sentences. Condition (2) is a weak form of the principle of compositionality.

The semantics we defined in Chapter 4 satisfy both conditions. If ψ is an IF formula, and U is a finite set of variables that contains Free(ψ), let $\|\psi\|^+_{\mathbb{M},U}$ denote the set of winning assignment teams for ψ in $\mathbb{M}$ with domain U, and let $\|\psi\|^-_{\mathbb{M},U}$ denote the set of losing assignment teams with domain U. We take $\|\psi\|_{\mathbb{M},U}$ to be an ordered pair whose first coordinate is $\|\psi\|^+_{\mathbb{M},U}$ and whose second coordinate is $\|\psi\|^-_{\mathbb{M},U}$. For every finite set of variables U, define TRUE$_{\mathbb{M},U}$ to be a pair whose coordinates are

$$\text{TRUE}^+_{\mathbb{M},U} = \mathscr{P}(M^U) \quad \text{and} \quad \text{TRUE}^-_{\mathbb{M},U} = \{\varnothing\},$$

respectively. In particular, note that $\text{TRUE}^+_{\mathbb{M}} = \mathscr{P}(\{\varnothing\})$.

Theorem 8.2 $\| \cdot \|$ *is an adequate semantics for* IF *logic.*

Proof (1) Let φ be an IF sentence, and let $\mathbb{M}$ be a suitable structure. Then $\mathbb{M} \models^+_{\text{GTS}} \varphi$ if and only if $\{\varnothing\}$ is a winning team of assignments for φ. It follows from Proposition 5.2 that

$$\|\varphi\|^+_\mathbb{M} = \mathscr{P}(\{\varnothing\}) \quad \text{and} \quad \|\varphi\|^-_\mathbb{M} = \{\varnothing\}.$$

Hence $\|\varphi\|_\mathbb{M} = \text{TRUE}$.

(2) Suppose ψ is a subformula of φ, and $\|\psi\|_{\mathbb{M}, A_\varphi(\psi)} = \|\psi'\|_{\mathbb{M}, A_\varphi(\psi)}$. If φ' is the result of replacing ψ by ψ', then by mimicking the proof of Proposition 5.17 one can easily show that $\|\varphi\|_\mathbb{M} = \|\varphi'\|_\mathbb{M}$. $\dashv$

The semantics defined in Chapter 4 are also *fully abstract* in the sense that whenever two IF formulas do not have the same meaning, there is a structure in which replacing one formula for the other changes the truth value of an IF sentence.

Theorem 8.3 (Hodges [32, Theorem 7.6]) *Let U be a finite set of variables, and let ψ and ψ' be* IF *formulas whose free variables are contained in U. Suppose that for every* IF *sentence φ in which ψ is a subformula and $A_\varphi(\psi) = U$, we have $\varphi \equiv \varphi'$ whenever φ' is the result of replacing ψ by ψ' in φ. Then $\psi \equiv_U \psi'$.*

Proof Suppose there is a suitable structure $\mathbb{M}$ such that

$$\|\psi\|_{\mathbb{M}, U} \neq \|\psi'\|_{\mathbb{M}, U} .$$

Then without loss of generality there is an an assignment team $X \subseteq M^U$ such that $X \in \|\psi\|^+_{\mathbb{M}, U}$ but $X \notin \|\psi'\|^+_{\mathbb{M}, U}$.

Let $U = \{x_1, \ldots, x_n\}$, let R be a fresh n-ary relation symbol, and let $\mathbb{M}^*$ be the expansion of $\mathbb{M}$ in which

$$R^{\mathbb{M}^*} = \Big\{ \big(s(x_1), \ldots, s(x_n)\big) \colon s \in X \Big\}.$$

Let φ be the IF sentence

$$\forall x_1 \ldots \forall x_n \big(\neg R(x_1, \ldots, x_n) \vee \psi\big).$$

Then $\mathbb{M}^* \models^+ \varphi$ because

$$\mathbb{M}^*, \big(M^U - X\big) \models^+ \neg R(x_1, \ldots, x_n) \quad \text{and} \quad \mathbb{M}^*, X \models^+ \psi,$$

whereas $\mathbb{M}^* \not\models^+ \varphi'$ because if it did we would have $\mathbb{M}^*, X \models^+ \psi'$ which implies $\mathbb{M}, X \models^+ \psi'$, contrary to hypothesis. $\dashv$

Now we will show that there is no adequate semantics $|\cdot|$ for IF logic such that for every IF formula φ whose free variables are contained in U we have $|\varphi|_{\mathrm{M},U} \subseteq M^U$. The proof is a simple counting argument based on the fact that any adequate semantics for IF logic must have more distinct meanings than there are subsets of M^U.

Definition 8.4 (Cameron and Hodges [11])   Let M be a structure and U a finite set of variables. An (M, U)-*suit* is a nonempty collection $\mathcal{S} \subseteq \mathscr{P}(M^U)$ of assignment teams with domain U that is downwards closed, i.e., $Y \subseteq X \in \mathcal{S}$ implies $Y \in \mathcal{S}$. A *double* (M, U)-*suit* is an ordered pair $(\mathcal{S}^+, \mathcal{S}^-)$ such that $\mathcal{S}^+$ and $\mathcal{S}^-$ are (M, U)-suits and $\mathcal{S}^+ \cap \mathcal{S}^- = \{\varnothing\}$.[2]   ⊣

It follows immediately from Propositions 5.1 and 5.2 that the meaning $\|\psi\|_{\mathrm{M},U}$ of an IF formula ψ in a suitable structure M relative to a finite set of variables U containing $\mathrm{Free}(\psi)$ is a double (M, U)-suit. Theorem 8.6 below shows this is the strongest possible characterization of meanings of IF formulas in finite structures.

Lemma 8.5   *Let M be a finite structure in which every element is named by a constant symbol, and let U be a finite set of variables. Then for every (M, U)-suit $\mathcal{S}$ there is an IF formula ψ such that $\mathcal{S} = \|\psi\|_{\mathrm{M},U}^+$.*

Proof   Let $U = \{x_1, \ldots, x_n\}$, and for every assignment $s \in M^U$ let ψ_s be the formula

$$x_1 = c_{s(x_1)} \wedge \ldots \wedge x_n = c_{s(x_n)},$$

where $c_{s(x_i)}$ is the constant symbol whose interpretation is $s(x_i)$. For every assignment team $X \subseteq M^U$, let ψ_X be the formula $\bigvee_{s \in X} \psi_s$. Observe that

$$\|\psi_X\|_{\mathrm{M},U}^+ = \mathscr{P}(X)$$
$$\|\psi_X\|_{\mathrm{M},U}^- = \mathscr{P}(M^U - X).$$

Also note that every (M, U)-suit $\mathcal{S}$ has the form

$$\mathcal{S} = \mathscr{P}(X_1) \cup \ldots \cup \mathscr{P}(X_\ell),$$

where each $X_k \subseteq M^U$ is a team of assignments with domain U. Let ψ be the IF formula

$$\psi_{X_1} \vee_{/U} \ldots \vee_{/U} \psi_{X_\ell}.$$

---

[2] Cameron and Hodges define a suit to be a collection of *nonempty* assignment teams that is closed under *nonempty* subsets. We allow the empty team of assignments to be both winning and losing for any IF formula, so we include it in our suits and double suits.

It follows from Lemma 5.12 that

$$\|\psi\|_{M,U}^{+} = \bigcup_{k=1}^{\ell} \|\psi_{X_k}\|_{M,U}^{+} = \mathcal{S}. \qquad \dashv$$

Since $\|\neg\varphi\|_{M,U}^{\pm} = \|\varphi\|_{M,U}^{\mp}$, it is also the case that for every (M, U)-suit $\mathcal{S}$ we can find an IF formula χ such that $\mathcal{S} = \|\chi\|_{M,U}^{-}$.

Theorem 8.6   *Let* M *be a finite structure in which every element is named by a constant symbol, and let* U *be a finite set of variables. If* M *has at least two elements, then every double* (M, U)-*suit* $(\mathcal{S}^{+}, \mathcal{S}^{-})$ *is the meaning of an IF formula in* M *relative to* U.

Proof   Let ψ and χ be IF formulas such that

$$\|\psi\|_{U}^{+} = \mathcal{S}^{+} \quad \text{and} \quad \|\chi\|_{U}^{-} = \mathcal{S}^{-}.$$

Let φ_{MP} be the Matching Pennies sentence $\forall x (\exists y / \{x\}) x = y$, where x and y are variables not in U. Let ψ' be the formula $\psi \vee_{/U} \varphi_{MP}$, and let χ' be $\chi \wedge_{/U} \varphi_{MP}$. Since $\|\varphi_{MP}\|_{M,U}^{\pm} = \{\varnothing\}$, we have

$$\|\psi'\|_{U}^{+} = \mathcal{S}^{+}, \qquad\qquad \|\chi'\|_{U}^{+} = \{\varnothing\},$$
$$\|\psi'\|_{U}^{-} = \{\varnothing\}, \qquad\qquad \|\chi'\|_{U}^{-} = \mathcal{S}^{-}.$$

Finally, let $X = \bigcup \mathcal{S}^{+}$. Then $\mathcal{S}^{+} \subseteq \mathscr{P}(X)$ and $\mathcal{S}^{-} \subseteq \mathscr{P}(M^{U} - X)$, so

$$\|\psi' \wedge_{/U} (\chi' \vee_{/U} \psi_X)\|_{U}^{+} = \mathcal{S}^{+},$$
$$\|\psi' \wedge_{/U} (\chi' \vee_{/U} \psi_X)\|_{U}^{-} = \mathcal{S}^{-}. \qquad \dashv$$

Thus, to obtain a measure of the expressiveness of IF formulas on finite models, one should count double suits. However, to prove the impossibility result we are aiming for, it is sufficient to only consider the truth coordinates of IF formulas. Thus we only need to count suits. Let $f(m, n)$ denote the number of (M, U)-suits when M is a structure of size m, and U is a set of n variables. Cameron and Hodges [11, §4] perform several calculations involving $f(m, n)$. In particular, they observe that when $n = 1$ there are as many (M, U)-suits as elements in the free distributive 1-lattice on m generators. Figure 8.1, which was adapted from [11, p. 679], shows the value of $f(m) = f(m, 1)$ compared to 2^m for $1 \leq m \leq 8$. One can see that $f(m)$ appears to grow much faster than 2^m, which is the crucial fact we will use to prove our desired impossibility result.

m	2^m	$f(m)$
1	2	2
2	4	5
3	8	19
4	16	167
5	32	7,580
6	64	7,828,353
7	128	2,414,682,040,997
8	256	56,130,437,228,687,557,907,787

Figure 8.1 Counting suits

Proposition 8.7 (Cameron and Hodges [11, Proposition 4.2(c)]) *If $m > 1$, then $f(m) > 2^m$.*

Proof Let $\mathbb{M}$ be a structure of size $m > 1$, and let $U = \{x\}$. For each $a \in M$, let $s_a = \{(x, a)\}$ be the assignment that assigns the value a to x. Then for any two distinct subsets $A, B \subseteq M$, the collections

$$\mathcal{S}_A = \{\{s_a\} \colon a \in A\} \cup \{\varnothing\},$$
$$\mathcal{S}_B = \{\{s_b\} \colon b \in B\} \cup \{\varnothing\},$$

consisting only of singleton assignment teams (and the empty assignment team) are distinct $(\mathbb{M}, \{x\})$-suits. Hence $2^m \leq f(m)$. If $a, b \in M$, then

$$\{\varnothing, \{s_a\}, \{s_b\}, \{s_a, s_b\}\}$$

is an $(\mathbb{M}, \{x\})$-suit that is distinct from every $\mathcal{S}_A$. Thus $2^m < f(m)$. ⊣

Theorem 8.8 (Cameron and Hodges [11, Theorem 6.1]) *If $|\cdot|$ is an adequate semantics for IF logic, then for each integer $m > 1$ there is a structure $\mathbb{M}$ of size m in which there are at least $f(m)$ distinct meanings $|\psi|_{\mathbb{M}, \{x\}}$ of IF formulas ψ with exactly one free variable.*

Proof Fix $m > 1$, and let $\mathbb{M}$ be a structure of size m in which every element is named by a constant symbol. Let

$$\big(\mathcal{S}_i : 0 \leq i < f(m) \big)$$

enumerate all the $(\mathbb{M}, \{x\})$-suits. By Lemma 8.5, for each i there is an IF formula ψ_i such that $\mathcal{S}_i = \|\psi_i\|_{\{x\}}^+$. Thus, for any two distinct $0 \leq i, j < f(m)$ there exists a nonempty assignment team $X_{ij} \subseteq M^{\{x\}}$ such that $\mathbb{M}, X_{ij} \models^+ \psi_i$ but $\mathbb{M}, X_{ij} \not\models^+ \psi_j$, or vice versa. Let χ_{ij} be an IF formula such that

$$\|\chi_{ij}\|_{\{x\}}^+ = \mathscr{P}(M^{\{x\}} - X_{ij}),$$

and let φ_{ij} be the IF sentence $\forall x(\psi_i \vee \chi_{ij})$. Then φ_{ij} has the property that its truth-value $\|\varphi_{ij}\|$ changes when ψ_i is replaced by ψ_j.

Now suppose $|\cdot|$ is a adequate semantics for IF logic. Since $|\cdot|$ agrees with $\|\cdot\|$ on sentences, the truth-value $|\varphi_{ij}|$ changes when we replace ψ_i by ψ_j. Hence $|\psi_i|_{\{x\}} \neq |\psi_j|_{\{x\}}$ by condition (2) of the definition of adequate semantics. Thus $|\cdot|$ assigns at least $f(m)$ distinct meanings in $\mathbb{M}$ to IF formulas with one free variable. ⊣

We conclude that there is no adequate semantics for IF logic for which the meaning of an IF formula is a set of assignments. In the single-variable case, there are only 2^m distinct sets of assignments in a structure of size m, while there is a structure of size m in which any adequate semantics must assign at least $f(m)$ distinct meanings to IF formulas with a single variable. By Proposition 8.7 we have $2^m < f(m)$ for any $m > 1$, which proves our claim.

8.2 IF modal logic

There is no algorithm to decide whether a given first-order sentence is satisfiable [12]. *A fortiori*, the satisfiability problem for IF logic is not decidable. In light of this negative result, it is worthwhile to study the computational virtues of fragments of IF logic. This will help us improve our understanding of independence-friendliness from a computational point of view. In this section we consider a fragment of IF logic that we call IF modal logic due to the resemblance it bears to the standard first-order translation of ordinary modal logic. We will see that this fragment cannot be translated to first-order logic, but that its satisfiability problem is decidable.

8.2.1 Syntax and semantics

Modal logic is tailored to graph-like structures $\mathbb{M}$ that interpret *modal vocabularies* $\{R, P_1, \ldots, P_n\}$ in which R is a binary relation symbol and $P_1, \ldots, P_n$ are predicate symbols. The elements of the universe are called *objects* or *states*. If $(a, b) \in R^{\mathbb{M}}$ then we say that state b is *accessible* from a or that b is a *successor* of a. The interpretation of each predicate symbol P_i is a subset of the universe consisting of those states that have the property $P_i^{\mathbb{M}}$.

The semantic game of a modal formula φ starts from a so-called *current state* or *current object* a. The quantifiers of modal logic ($\diamond$ and $\square$)

range over the successors of the current state, or game-theoretically, they prompt Eloise or Abelard to choose a successor of the current state. For instance, in the semantic game of the modal formula $\Diamond\varphi$ at the current state a, Eloise is to pick a successor b of a. The semantic game of φ then continues from b. Similarly, in the semantic game of the modal formula $\Box\varphi$ at a, Abelard is to pick a successor b of a.

For our purposes it will be convenient to regard modal logic as a fragment of first-order logic.[3]

Definition 8.9 Let L be a modal vocabulary and let x be a variable. The set ML_L^x is generated by the finite application of the following rules:

- If P is a predicate symbol in L, then $P(x) \in \mathrm{ML}_L^x$ and $\neg P(x) \in \mathrm{ML}_L^x$.
- If $\varphi, \varphi' \in \mathrm{ML}_L^x$, then $(\varphi \vee \varphi') \in \mathrm{ML}_L^x$ and $(\varphi \wedge \varphi') \in \mathrm{ML}_L^x$.
- If R is a binary relation symbol in L, y is a variable distinct from x, and $\varphi \in \mathrm{ML}_L^y$, then

$$\exists y\big(R(x,y) \wedge \varphi\big) \in \mathrm{ML}_L^x \quad \text{and} \quad \forall y\big(R(x,y) \to \varphi\big) \in \mathrm{ML}_L^x.$$

The modal language generated by the vocabulary L, denoted ML_L, is the union of ML_L^x for all variables x.   ⊣

The elements of ML_L are called ML_L *formulas*. A *modal formula* is an ML_L formula for some modal vocabulary L. When the vocabulary is irrelevant or clear from context we will not mention it explicitly.

Every ML_L formula is an FO_L formula for any modal vocabulary L. This allows us to transfer some technical machinery from first-order logic to modal logic, such as the notion of subformula and free variable. Every element in ML_L^x has precisely one free variable: x. The value of x is interpreted as the current state.

Unlike first-order quantifiers, modal quantifiers always occur bound by the relation R, e.g., $Qy\big(R(x,y)\ldots\big)$. This forces Eloise and Abelard to only pick successors of the current state.

We can also transfer the semantics of first-order logic to modal logic. Thus if φ is a modal formula with free variable x, $\mathbb{M}$ is a suitable structure, and s is an assignment whose domain includes x, then

$$\mathbb{M}, s \models \varphi$$

---

[3] It is well known that the current formulation is interchangeable with the customary syntax of modal logic in terms of propositional variables $p, q, \ldots$, negation $\neg$, connectives $\vee, \wedge$, and modalities $\Diamond, \Box$. The so-called "standard translation" mediates between the customary syntax of modal logic and its first-order rendering used here, see [5, p. 83].

expresses that Eloise has a winning strategy in the first-order semantic game $G(\mathbb{M}, s, \varphi)$.

Every modal formula is a first-order formula, but not every first-order formula is equivalent to a modal formula. Cases in point are

$$\exists y(x \neq y \wedge \varphi) \quad \text{and} \quad \exists y\big(R(y, x) \wedge \varphi\big)$$

for any modal formula φ with free variable y, see [5, p. 63]. Also, the formula

$$\exists z \forall y \Big(R(x, y) \rightarrow \big(R(y, z) \wedge \varphi\big)\Big)$$

has no equivalent in modal logic, for any modal formula φ with free variable z.

The good news is that modal logic is decidable [37]. An extensive literature has been developed on extensions and variants of modal logic and their computational properties, see [5, 41]. In a series of publications Tulenheimo, later joined by the third author, proposed an independence-friendly modal logic [52, 62–64].

Given the definition of modal logic, the definition of IF modal logic should come as no surprise.

Definition 8.10 Let L be a modal vocabulary, and let x be a variable. The set IFML_L^x is generated by the finite application of the following rules:

- If P is a predicate symbol in L, then $P(x) \in \text{IFML}_L^x$ and $\neg P(x) \in \text{IFML}_L^x$.
- If $\varphi, \varphi' \in \text{IFML}_L^x$, then $(\varphi \vee \varphi') \in \text{IFML}_L^x$ and $(\varphi \wedge \varphi') \in \text{IFML}_L^x$.
- If R is a binary relation symbol in L, y is a variable distinct from x, $\varphi \in \text{IFML}_L^y$, and W is a finite set of variables, then

$$(\exists y/W)\big(R(x, y) \wedge \varphi\big) \in \text{IFML}_L^x,$$
$$(\forall y/W)\big(R(x, y) \rightarrow \varphi\big) \in \text{IFML}_L^x.$$

The IF modal language generated by the modal vocabulary L, denoted IFML_L, is the union of IFML_L^x for all variables x.   ⊣

The elements of IFML_L are called IFML_L *formulas*, while an IF *modal formula* is an IFML_L formula for some vocabulary L. Every IF modal formula is an IF formula, and in every IF modal formula $(Qy/W)\big(R(x, y) \ldots\big) \in \text{IFML}_L^x$, the variable x is free.

As before we can transfer the semantic apparatus of IF logic to IF modal logic. Thus if φ is an IF modal formula, $\mathbb{M}$ is a suitable structure, and X is an assignment team whose domain contains Free(φ), then

$$\mathbb{M}, X \models^{\pm} \varphi$$

expresses that Eloise/Abelard has a winning strategy for $G(\mathbb{M}, X, \varphi)$.

8.2.2 Model theory

Tulenheimo [62] showed that there are IF modal formulas that are not truth equivalent relative to $\{x\}$ to any ordinary modal formula. The simplest case in point is

$$\forall y \Big(R(x,y) \to (\exists z/\{y\}) \big(R(y,z) \wedge \varphi \big) \Big).$$

By Theorem 5.35 and Proposition 6.22(c), this formula is truth equivalent relative to $\{x\}$ to the IF formula

$$\exists z \forall y \Big(R(x,y) \to \big(R(y,z) \wedge \varphi \big) \Big),$$

which, as we saw earlier, has no equivalent in modal logic.

Later it was proved that there are also IF modal formulas that are not truth equivalent to any first-order formula [64]. Therefore, the notion of independence-friendliness is powerful enough to increase the expressive power of modal logic beyond first-order logic.

Theorem 8.11 *There is an IFML$_L$ formula that is not equivalent to any FO$_L$ formula.*

Proof The proof involves the class of *seastar* structures. A seastar structure is a directed graph whose edges radiate out from a central point. For each $1 \le m \le 4$, the set of vertices exactly m steps from the center is called the *ring* of radius m. Every ring in a seastar structure has the same number of vertices. The *circumference* of a seastar structure is the number of nodes in each of its rings, and $\mathbb{S}_n$ denotes the seastar structure with circumference n. The seastar structure with circumference $n = 8$ is drawn in Figure 8.2. Let a denote the middle object in $\mathbb{S}_n$, and let $s = \{(x,a)\}$ be the assignment that assigns a to the variable x.

The proof comes in two parts. First we show that for every first-order formula φ with free variable x there is an n such that $\mathbb{S}_n$ and $\mathbb{S}_{n+1}$ are indistinguishable, that is,

$$\mathbb{S}_n, s \models \varphi \quad \text{implies} \quad \mathbb{S}_{n+1}, s \models \varphi.$$

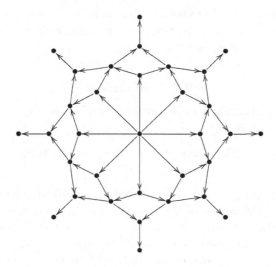

Figure 8.2 The seastar structure $\mathbb{S}_n$ with circumference $n = 8$.

This can be proved by means of Ehrenfeucht-Fraïssé games [18, 20]. In the proof we can exploit the resemblance of $\mathbb{S}_n$ with the cyclic graph with $2n$ objects. It is well known that first-order logic cannot define evenness on cyclic graphs (see [16, p. 23]), that is, it cannot distinguish cyclic graphs of size $2n$ from cyclic graphs of size $2n + 1$.

Second we show that the IF modal formula $\forall y\big(R(x, y) \rightarrow (\psi \vee \psi)\big)$, where ψ is

$$\forall u\Big(R(y, u) \rightarrow (\exists v/\{u\})\,[R(u, v) \wedge (\exists w/\{y, v\})\,R(v, w)]\Big)$$

has the property that n is even if and only if $\mathbb{S}_n, s \models^+ \varphi$. Using trump semantics, $\mathbb{S}_n, s \models^+ \varphi$ if and only if $\mathbb{S}_n, \{s\}[y, A] \models^+ \psi \vee \psi$, where A is the set of objects in the inner ring of $\mathbb{S}_n$, that is, the objects that are accessible from its middle object a.

If n is even, there is an *exclusive cover* of A, that is, a cover $B \cup B'$ of A such that for both $C \in \{B, B'\}$, no two objects in C have a shared successor. Let $B \cup B'$ be an exclusive cover of A. It can be shown that $\mathbb{S}_n, \{s\}[y, B] \models^+ \psi$ and $\mathbb{S}_n, \{s\}[y, B'] \models^+ \psi$.

If n is odd, there is no exclusive cover. Consequently, for every cover $B \cup B'$ of A there is a $C \in \{B, B'\}$ that contains two objects with a shared successor. One can prove that for this C we have $\mathbb{S}_n, \{s\}[y, C] \not\models^+ \psi$. Hence $\mathbb{S}_n, \{s\}[y, A] \not\models^+ \psi \vee \psi$. $\dashv$

Observe that the IF modal formula used in the second part of the proof, the one that has no first-order equivalent, lacks perfect recall for Eloise (compare Corollary 6.24).

8.2.3 Decidability

In a recent publication by the third author [53] it was shown that IF modal logic is decidable.

Theorem 8.12 *The class of satisfiable, regular IFML$_L$ formulas is decidable.*

Proof The proof given in [53] is based on a tableau argument. It revolves around the notion of "witness system" that collects the satisfiability constraints of each subformula ψ in a regular IF modal formula φ. Technically a witness system (w, S) consists of two objects. The first object w is a function that maps each subformula ψ to a team of assignments A^U, where U is the set of variables that appear free in the subformulas in φ that contain ψ as a subformula, and A is a sufficiently large set of objects. The second object S is a binary relation on A.

The idea is that $w(\psi)$ is a team of assignments that suffices to satisfy ψ. Thus, if (w, S) is a witness system of φ, we require that for every subformula ψ of the form $\chi \vee \chi'$ in φ,

$$w(\psi) = w(\chi) \cup w(\chi)$$

and similarly we require that for every subformula ψ of the form $\chi \wedge \chi'$,

$$w(\psi) = w(\chi) = w(\chi').$$

Each assignment $s \in w(\psi)$ with domain $y_1, \ldots, y_n$ introduces the objects $s(y_1), \ldots, s(y_n)$. The size of a witness system refers to the total number of distinct objects introduced by its assignments. The relation S contains all pairs of objects thus introduced that need to be connected to satisfy the subformulas in φ. If (w, S) is a witness system of φ, we require that for every subformula ψ of the form $(\exists y/W)(R(x, y) \wedge \psi')$ in φ there is a W-uniform function f with domain $w(\psi)$ such that

$$w(\psi)[y, f] = w(\psi')$$

and $(s(x), f(s)) \in S$ for every $s \in w(\psi)$.

If (w, S) is a witness system of φ, we require that for every subformula ψ of the form $(\forall y/W)(R(x, y) \rightarrow \psi')$,

$$w(\psi)[y, B] = w(\psi'),$$

where $B = \{b \in A : (s(x), b) \in S$ for some $s \in w(\psi)\}$. So the universal quantifier only affects objects that are reachable from the objects in B.

In order to avoid the all-empty witness system, we require $w(\varphi)$ to be nonempty if (w, S) is a witness system of φ.

For literals ψ of the form $P(x)$, the set $\{s(x) : s \in w(\psi)\}$ contains the objects that should be P objects, whereas for literals ψ' of the form $\neg P(y)$, the set $\{s(y) : s \in w(\psi')\}$ contains the objects that should not be P objects. There is an inconsistency in the witness system if there are two such subformulas ψ and ψ' such that $\{s(x) : s \in w(\psi)\}$ and $\{s(y) : s \in w(\psi')\}$ overlap. In that case we say that (w, S) is *closed*. It it is not closed, we say it is *open*.

We can now prove that φ is satisfiable if and only if φ has an open witness system. The direction from right-to-left is straightforward, as we can easily convert a witness system to a structure $\mathbb{M}$ and prove that $\mathbb{M}, w(\psi) \models^+ \psi$, for every subformula ψ in φ, including φ itself. The converse direction requires us to distill an open witness system from a structure-assignment pair that is known to satisfy φ by carefully managing the order in which we construct the sets $w(\psi)$ for the subformulas ψ.

Finally it can be shown that there is a finite upper bound on the size of the witness system of φ. A naive algorithm would thus iterate through all witness systems. It accepts φ and terminates the moment it finds an open witness system of φ . If it has checked all witness systems that have size less than the upper bound for φ, then it rejects φ. This proves the theorem.

The regularity constraint is an artifact of the way the language was presented in [53]. It is conceivable that this constraint can be lifted without affecting the decidability of IF modal logic. $\dashv$

References

[1] P. Abelard and Héloïse. *The Letters of Abelard and Eloise.* Penguin, London, revised edition, 2003. Translated by Betty Radice.

[2] J. F. A. K. van Benthem. Logic games are complete for game logics. *Studia Logica,* 75:183–203, 2003.

[3] J. F. A. K. van Benthem. Probabilistic features in logic games. In D. Kolak and D. Symons, editors, *Quantifiers, Questions and Quantum Physics,* pages 189–194. Springer, Dordrecht, 2004.

[4] J. F. A. K. van Benthem. The epistemic logic of IF games. In R. E. Auxier and L. E. Hahn, editors, *The Philosophy of Jaakko Hintikka,* volume 30 of *Library of Living Philosophers,* pages 481–513. Open Court, Chicago, 2006.

[5] P. Blackburn, M. de Rijke, and Y. Venema. *Modal Logic.* Cambridge University Press, Cambridge, UK, 2001.

[6] P. Blackburn, J. F. A. K. van Benthem, and F. Wolter, editors. *Handbook of Modal Logic,* volume 3 of *Studies in Logic and Practical Reasoning.* Elsevier, Amsterdam, 2007.

[7] A. Blass and Y. Gurevich. Henkin quantifiers and complete problems. *Annals of Pure and Applied Logic,* 32:1–16, 1986.

[8] J. P. Burgess. A remark on Henkin sentences and their contraries. *Notre Dame Journal of Formal Logic,* 44:185–188, 2003.

[9] X. Caicedo, F. Dechesne, and T. M. V. Janssen. Equivalence and quantifier rules for logic with imperfect information. *Logic Journal of IGPL,* 17:91–129, 2009.

[10] X. Caicedo and M. Krynicki. Quantifiers for reasoning with imperfect information and Σ_1^1-logic. In W. A. Carnielli and I. M. Ottaviano, editors, *Advances in Contemporary Logic and Computer Science: Proceedings of the Eleveth Brazilian Conference on Mathematical Logic, May 6-10, 1996,* volume 235 of *Contemporary Mathematics,* pages 17–31. American Mathematical Society, 1999.

[11] P. J. Cameron and W. Hodges. Some combinatorics of imperfect information. *Journal of Symbolic Logic,* 66:673–684, 2001.

[12] A. Church. A note on the Entscheidungsproblem. *Journal of Symbolic Logic,* 1:40–41, 1936.

[13] W. Craig. Three uses of the Herbrand-Gentzen theorem in relating model theory and proof theory. *Journal of Symbolic Logic*, 22:269–285, 1957.

[14] F. Dechesne. *Game, Set, Maths: Formal investigations into logic with imperfect information.* PhD thesis, Tilburg University, Tilburg, 2005.

[15] R. Dedekind. *Was sind und was sollen die Zahlen?* Braunschweig, Germany, 1888.

[16] H.-D. Ebbinghaus and J. Flum. *Finite Model Theory.* Springer-Verlag, Berlin, 1999.

[17] H.-D. Ebbinghaus, J. Flum, and W. Thomas. *Mathematical Logic.* Undergraduate Texts in Mathematics. Springer-Verlag, Berlin, 2nd edition, 1994.

[18] A. Ehrenfeucht. An application of games to the completeness problem for formalized theories. *Fundamenta Mathematicae*, 49:129–141, 1961.

[19] H. B. Enderton. Finite partially ordered quantifiers. *Zeitschrift für Mathematische Logik und Grundlagen der Mathematik*, 16:393–397, 1970.

[20] R. Fraïssé. Sur quelques classifications des systèmes de relations. Publications scientifiques, série A, 35–182, Université d'Alger, 1954.

[21] D. Gale and F. Stewart. Infinite games with perfect information. In H. W. Kuhn and A. W. Tucker, editors, *Contributions to the Theory of Games II*, volume 28 of *Annals of Mathematics Studies*, pages 245–266. Princeton University Press, Princeton, 1953.

[22] P. Galliani. Game values and equilibria for undetermined sentences of dependence logic. Master's thesis. Master of Logic Series 2008-08, University of Amsterdam, Amsterdam, 2008.

[23] P. Galliani and A. L. Mann. Lottery semantics. In J. Kontinen and J. Väänänen, editors, *Proceedings of the ESSLLI Workshop on Dependence and Independence in Logic*, pages 30–54, Copenhagen, August 16–20, 2010.

[24] L. Hella and G. Sandu. Partially ordered connectives and finite graphs. In M. Krynicki, M. Mostowski, and L. W. Szczerba, editors, *Quantifiers: Logics, Models and Computation*, volume 2, pages 79–88. Kluwer Academic Publishers, Dordrecht, 1995.

[25] L. Henkin. Some remarks on infinitely long formulas. In P. Bernays, editor, *Infinitistic Methods: Proceedings of the Symposium on Foundations of Mathematics, Warsaw, 2–9 September 1959*, pages 167–183, Oxford, 1961. Pergamon Press.

[26] J. Hintikka. Language-games for quantifiers. In *Studies in Logical Theory*, volume 2 of *American Philosophical Quarterly Monograph Series*, pages 46–72. Basil Blackwell, Oxford, 1968.

[27] J. Hintikka. Quantifiers vs. quantification theory. *Dialectica*, 27:329–358, 1973.

[28] J. Hintikka. *Principles of Mathematics Revisited.* Cambridge University Press, Cambridge, UK, 1996.

[29] J. Hintikka and J. Kulas. *The Game of Language.* Reidel, Dordrecht, 1983.

[30] J. Hintikka and G. Sandu. Informational independence as a semantic phenomenon. In J. E. Fenstad *et al.*, editors, *Logic, Methodology and Philosophy of Science*, volume 8, pages 571–589. Elsevier, Amsterdam, 1989.

[31] J. Hintikka and G. Sandu. Game-theoretical semantics. In J. F. A. K. van Benthem and A. ter Meulen, editors, *Handbook of Logic and Language*, pages 361–481. North Holland, Amsterdam, 1997.

[32] W. Hodges. Compositional semantics for a language of imperfect information. *Logic Journal of the IGPL*, 5:539–563, 1997.

[33] W. Hodges. Some strange quantifiers. In J. Mycielski, G. Rozenberg, and A. Salomaa, editors, *Structures in Logic and Computer Science: A Selection of Essays in Honor of A. Ehrenfeucht*, Lecture Notes in Computer Science, pages 51–65. Springer-Verlag, 1997.

[34] S. C. Kleene. *Introduction to Metamathematics*. Van Nostrand, 1952.

[35] H. W. Kuhn. Extensive games. *Proceedings of the National Academy of Sciences of the United States of America*, 36:570–576, 1950.

[36] H. W. Kuhn. Extensive games and the problem of information. In H. W. Kuhn and A. W. Tucker, editors, *Contributions to the Theory of Games II*, volume 28 of *Annals of Mathematics Studies*, pages 193–216. Princeton University Press, Princeton, 1953.

[37] R. E. Ladner. The computational complexity of provability in systems of modal propositional logic. *SIAM Journal on Computing*, 6:467–480, 1977.

[38] D. K. Lewis. *Convention: A Philosophical Study*. Harvard University Press, Cambridge, Massachusetts, 1969.

[39] A. L. Mann. *Independence-Friendly Cylindric Set Algebras*. PhD thesis, University of Colorado at Boulder, 2007.

[40] A. L. Mann. Independence-friendly cylindric set algebras. *Logic Journal of IGPL*, 17:719–754, 2009.

[41] M. Marx. Complexity of modal logic. In Blackburn *et al.* [6], pages 139–179.

[42] J. Nash. Non-cooperative games. *Annals of Mathematics*, 54:286–295, 1951.

[43] J. von Neumann. Zur Theorie der Gesellschaftsspiele. *Mathematische Annalen*, 100:295–320, 1928.

[44] M. J. Osborne. *An Introduction to Game Theory*. Oxford University Press, Oxford, 2003.

[45] M. J. Osborne and A. Rubinstein. *A Course in Game Theory*. MIT Press, Cambridge, Massachusetts, 1994.

[46] C. S. Peirce. *Reasoning and the Logic of Things*. Harvard University Press, Cambridge, Massachusetts, 1992.

[47] M. Piccione and A. Rubinstein. On the interpretation of decision problems with imperfect recall. *Games and Economic Behavior*, 20:3–24, 1997.

[48] T. E. S. Raghavan. Zero-sum two person games. In R. J. Aumann and S. Hart, editors, *Handbook of Game Theory with Economic Applications*, volume 2, pages 736–759. Elsevier, Amsterdam, 1994.

[49] G. Sandu. On the logic of informational independence and its applications. *Journal of Philosophical Logic*, 22:29–60, 1993.

[50] G. Sandu. The logic of informational independence and finite models. *Logic Journal of the IGPL*, 5:79–95, 1997.

[51] G. Sandu and J. Väänänen. Partially ordered connectives. *Mathematical Logic Quarterly*, 38:361–372, 1992.

[52] M. Sevenster. *Branches of Imperfect Information: Logic, Games, and Computation*. PhD thesis, University of Amsterdam, Amsterdam, 2006.

[53] M. Sevenster. Decidability of independence-friendly modal logic. *Review of Symbolic Logic*, 3:415–441, 2010.

[54] M. Sevenster and G. Sandu. Equilibrium semantics of languages of imperfect information. *Annals of Pure and Applied Logic*, 161:618–631, 2010.

[55] T. Skolem. Logisch-kombinatorische Untersuchungen über die Erfüllbarkeit oder Beweisbarkeit mathematischer Sätze nebst einem Theoreme über dichte Mengen. *Videnskapsselskapet Skrifter, I. Matematisk-naturvidenskabelig Klasse*, 4:1–36, 1920.

[56] T. Skolem. Logico-combinatorial investigations in the satisfiability or provability of mathematical propositions: A simplified proof of a theorem by L. Löwenheim and generalizations of the theorem. In J. van Heijenoort, editor, *From Frege to Gödel: A Source Book in Mathematical Logic*, pages 254–263. Harvard University Press, Cambridge, Massachusetts, 1967.

[57] T. Skolem. *Selected Works in Logic*. Scandinavian University Press, Oslo, 1970.

[58] A. Tarski. *Pojęcie prawdy w językach nauk dedukcyjnych (On the concept of truth in languages of deductive sciences)*. Warsaw, 1933. English translation in (Tarski 1983), pages 152–278.

[59] A. Tarski. The semantic conception of truth and the foundations of semantics. *Philosophy and Phenomenological Research*, 4:341–376, 1944.

[60] A. Tarski. *Logic, Semantics, Metamathematics: Papers from 1923 to 1938*. Hackett, Indianapolis, 2nd edition, 1983.

[61] A. Tarski and R. L. Vaught. Arithmetical extensions of relational systems. *Compositio Mathematica*, 13:81–102, 1956–1958.

[62] T. Tulenheimo. On IF modal logic and its expressive power. In P. Balbiani, N.-Y. Suzuki, F. Wolter, and M. Zakharyaschev, editors, *Advances in Modal Logic*, volume 4, pages 474–498. King's College Publications, 2003.

[63] T. Tulenheimo. *Independence-Friendly Modal Logic*. PhD thesis, University of Helsinki, Helsinki, 2004.

[64] T. Tulenheimo and M. Sevenster. On modal logic, IF logic and IF modal logic. In I. Hodkinson and Y. Venema, editors, *Advances in Modal Logic*, volume 6, pages 481–501. College Publications, 2006.

[65] J. Väänänen. A remark on nondeterminacy in IF logic. In T. Aho and A.-V. Pietarinen, editors, *Truth and Games: Essays in Honour of Gabriel Sandu*, chapter 4, pages 71–77. Societas Philosophica Fennica, Helsinki, 2006.

[66] J. Väänänen. *Dependence Logic*. Cambridge University Press, Cambridge, UK, 2007.

[67] W. Walkoe. Finite partially-ordered quantification. *Journal of Symbolic Logic*, 35:535–555, 1970.

[68] L. Wittgenstein. *Philosophical Investigations*. Basil Blackwell, Oxford, 1958. Translated by G. E. M. Anscombe.

Index

Printed in the United States
by Baker & Taylor Publisher Services